Paleomagnetic Correlation and Ages of Basalt Flow Groups in Coreholes At and Near the Naval Reactors Facility, Idaho National Laboratory, Idaho

By Duane E. Champion, Linda C. Davis, Mary K.V. Hodges, and Marvin A. Lanphere

DOE/ID-22223
Prepared in cooperation with the U.S. Department of Energy

Scientific Investigations Report 2013–5012

U.S. Department of the Interior
U.S. Geological Survey

U.S. Department of the Interior
KEN SALAZAR, Secretary

U.S. Geological Survey
Suzette M. Kimball, Acting Director

U.S. Geological Survey, Reston, Virginia: 2013

Contents

Contents—Continued

Plate

Figures

Tables

Conversion Factors, Datums, and Abbreviations and Acronyms

Conversion Factors

SI to Inch-Pound

Multiply	By	To obtain
centimeter (cm)	0.3937	inch (in.)
cubic kilometer (km^3)	0.2399	cubic mile (mi^3)
cubic meter (m^3)	35.3145	cubic foot (ft^3)
kilometer (km)	0.6214	mile (mi)
meter (m)	3.2808	foot (ft)
square kilometer (km^2)	0.3861	square mile (mi^2)

Temperature in degrees Celsius (°C) may be converted to degrees Fahrenheit (°F) as follows:

$$°F= (1.8×°C) +32.$$

Inch-Pound to SI

Multiply	By	To obtain
cubic mile (mi^3)	4.168	cubic kilometer (km^3)
foot (ft)	0.3048	meter (m)
mile (mi)	1.609	kilometer (km)
square mile (mi^2)	2.590	square kilometer (km^2)

Temperature in degrees Fahrenheit (°F) may be converted to degrees Celsius (°C) as follows:

$$°C=(°F-32)/1.8.$$

Datums

Vertical coordinate information is referenced to National Geodetic Vertical Datum of 1929 (NGVD 29).

Horizontal coordinate information is referenced to the North American Datum of 1927 (NAD 27).

Altitude, as used in this report, refers to distance above the vertical datum.

Abbreviations and Acronyms

Abbreviation or Acronym	Definition
AF	Alternating field
ATRC	Advanced Test Reactor Complex (formerly known as RTC, Reactor Technology Complex, and TRA, Test Reactor Area)
CFA	Central Facilities Area
DOE	U.S. Department of Energy
ESRP	Eastern Snake River Plain
ICPP	Idaho Chemical Processing Plant
INEL	Idaho National Engineering Laboratory (1974–97)
INEEL	Idaho National Engineering and Environmental Laboratory (1997–2005)
INL	Idaho National Laboratory (2005-present)
INTEC	Idaho Nuclear Technology and Engineering Center
IRM	Isothermal remanent magnetization
ka	Thousand years old
k.y.	Thousand years
Ma	Million years old
NRF	Naval Reactors Facility
NRTS	National Reactor Testing Station (1949–74)
RWMC	Radioactive Waste Management Complex
TAN	Test Area North
USGS	U.S. Geological Survey

Paleomagnetic Correlation and Ages of Basalt Flow Groups in Coreholes at and near the Naval Reactors Facility, Idaho National Laboratory, Idaho

By Duane E. Champion, Linda C. Davis, Mary K.V. Hodges, and Marvin A. Lanphere

Abstract

Paleomagnetic inclination and polarity studies were conducted on subcore samples from eight coreholes located at and near the Naval Reactors Facility (NRF), Idaho National Laboratory (INL). These studies were used to characterize and to correlate successive stratigraphic basalt flow groups in each corehole to basalt flow groups with similar paleomagnetic inclinations in adjacent coreholes. Results were used to extend the subsurface geologic framework at the INL previously derived from paleomagnetic data for south INL coreholes. Geologic framework studies are used in conceptual and numerical models of groundwater flow and contaminant transport. Sample handling and demagnetization protocols are described, as well as the paleomagnetic data averaging process.

Paleomagnetic inclination comparisons among NRF coreholes show comparable stratigraphic successions of mean inclination values over tens to hundreds of meters of depth. Corehole USGS 133 is more than 5 kilometers from the nearest NRF area corehole, and the mean inclination values of basalt flow groups in that corehole are somewhat less consistent than with NRF area basalt flow groups. Some basalt flow groups in USGS 133 are missing, additional basalt flow groups are present, or the basalt flow groups are at depths different from those of NRF area coreholes.

Age experiments on young, low potassium olivine tholeiite basalts may yield inconclusive results; paleomagnetic and stratigraphic data were used to choose the most reasonable ages. Results of age experiments using conventional potassium argon and argon-40/argon-39 protocols indicate that the youngest and uppermost basalt flow group in the NRF area is 303 ± 30 ka and that the oldest and deepest basalt flow group analyzed is 884 ± 53 ka.

A south to north line of cross-section drawn through the NRF coreholes shows corehole-to-corehole basalt flow group correlations derived from the paleomagnetic inclination data. From stratigraphic top to bottom, key results include the following:

- The West of Advanced Test Reactor Complex (ATRC) flow group is the uppermost basalt flow group in the NRF area and correlates among seven continuously cored holes in this study under surficial sediments. The West of ATRC flow group is also found in coreholes near the ATRC, the Idaho Nuclear Technology and Engineering Center (INTEC), and in corehole USGS 129.

- The ATRC Unknown Vent flow group correlates among seven continuously cored holes in this study underlying the West of ATRC flow group and a sedimentary interbed. Additional paleomagnetic inclination and stratigraphic data derived from the NRF coreholes changed the previously reported interpretation of the subsurface distribution of this basalt flow group. The ATRC Unknown Vent flow group also is found in coreholes near the ATRC and INTEC.

- The Central Facilities Area (CFA) Buried Vent flow group correlates among all eight coreholes in the NRF area. It also is found in coreholes near the CFA and the Radioactive Waste Management Complex (RWMC) to the south. This basalt flow group is thickest near the CFA, which may indicate proximity to the vent. The State Butte flow group is found below the CFA Buried Vent flow group in the four northern NRF coreholes. It correlates to the State Butte surface vent located just northeast of the NRF. It is not found in coreholes south of the NRF.

- The Atomic Energy Commission (AEC) Butte flow group is found in coreholes USGS 133, NRF 6P, and NRF 7P. It probably underlies coreholes NRF B18-1, NRF 89-05, and NRF 89-04, but those coreholes were not drilled deeply enough to penetrate the flow group. The AEC Butte flow group vent is exposed at the surface near the ATRC, and its flows are found in many coreholes near the ATRC and INTEC. The AEC Butte flow group abruptly pinches out against the Matuyama Chron reversed polarity flows of the East Matuyama Middle flow group between coreholes NRF 7P and NRF 15.

- The East Matuyama Middle flow group correlates between coreholes NRF 15 and NRF 16 and may correlate to coreholes NPR Test/W-02 and ANL-OBS-A-001.

- The North Late Matuyama flow group correlates among coreholes USGS 133, NRF 6P, NRF 7P, NRF 15, and NRF 16. It probably underlies coreholes NRF B18-1, NRF 89-05, and NRF 89-04, but those coreholes were not drilled deeply enough to penetrate the flow group. The vent that produced the North Late Matuyama flow group may be located in the general NRF area because it is thickest near corehole NRF 6P.

- The Matuyama flow group is found in coreholes in the southern INL from south of the RWMC to corehole USGS 133 and may extend north to corehole NRF 15. The Matuyama flow group is thickest near the RWMC and thins to the north.

- The Jaramillo (Matuyama) flow group is found in corehole NRF 15, which is the deepest NRF corehole, and shows that the basalt flow group is thick in the subsurface at NRF. This flow group is thickest between the RWMC and INTEC and thins towards the ATRC and NRF.

Introduction

The U.S. Atomic Energy Commission [(AEC), now the U.S. Department of Energy (DOE)] established the National Reactor Testing Station (NRTS) on about 2,300 km^2 of the eastern Snake River Plain (ESRP) in southeastern Idaho in 1949. The NRTS was established to develop peacetime atomic energy, nuclear safety research, defense programs, and advanced energy concepts. The name of the laboratory has been changed to reflect changes in the research focus of the laboratory. Names formerly used for the laboratory, from earliest to most recent, were the National Reactor Testing Station [(NRTS) 1949–74], the Idaho National Engineering Laboratory [(INEL) 1974–97], and the Idaho National Engineering and Environmental Laboratory [(INEEL) 1997–2005]. Since 2005, the laboratory has been known as the Idaho National Laboratory (INL) (fig. 1).

U.S. Geological Survey (USGS) scientists have been studying the geology, petrography, paleomagnetism, age of basalt lava flows, and hydrology of the ESRP for more than 100 years beginning with Russell (1902). Studies of the geologic framework of the ESRP at and near the INL intensified in 1949 when feasibility studies for siting of the NRTS began. Studies included evaluation of hydraulic properties of the aquifer, seismic and volcanic hazards, facility design and construction, and the evolution of basaltic volcanism on the ESRP.

Wastewater that contained chemical and radiochemical wastes was discharged to ponds and wells, and solid and liquid wastes were buried in trenches and pits excavated in surficial sediments at the INL. Some wastewater continues to be disposed to infiltration and evaporation ponds. Concern about subsurface movement of contaminants from these wastes increased the number and variety of studies of subsurface geology and hydrology to provide information for conceptual and numerical models of groundwater flow and contaminant transport (Anderson and Lewis, 1989; Anderson, 1991; Anderson and Bowers, 1995; Anderson and Liszewski, 1997; Anderson and others, 1996a, 1996b; Ackerman and others, 2006, 2010).

In 1974, studies began to evaluate potential volcanic hazards at the INL. Subsequently, these studies were expanded in an evaluation of subsurface stratigraphy as it may have influenced hydrologic properties of the saturated and unsaturated zones of the ESRP aquifer. These studies have included geologic mapping and various studies of cored and surface basalt flows such as petrologic and paleomagnetic investigations and radiometric age measurements. Many investigations were carried out on selected drill cores from different facilities at the INL, such as the Radioactive Waste Management Complex (RWMC), the Idaho Nuclear Technology and Engineering Center (INTEC), and Test Area North (TAN).

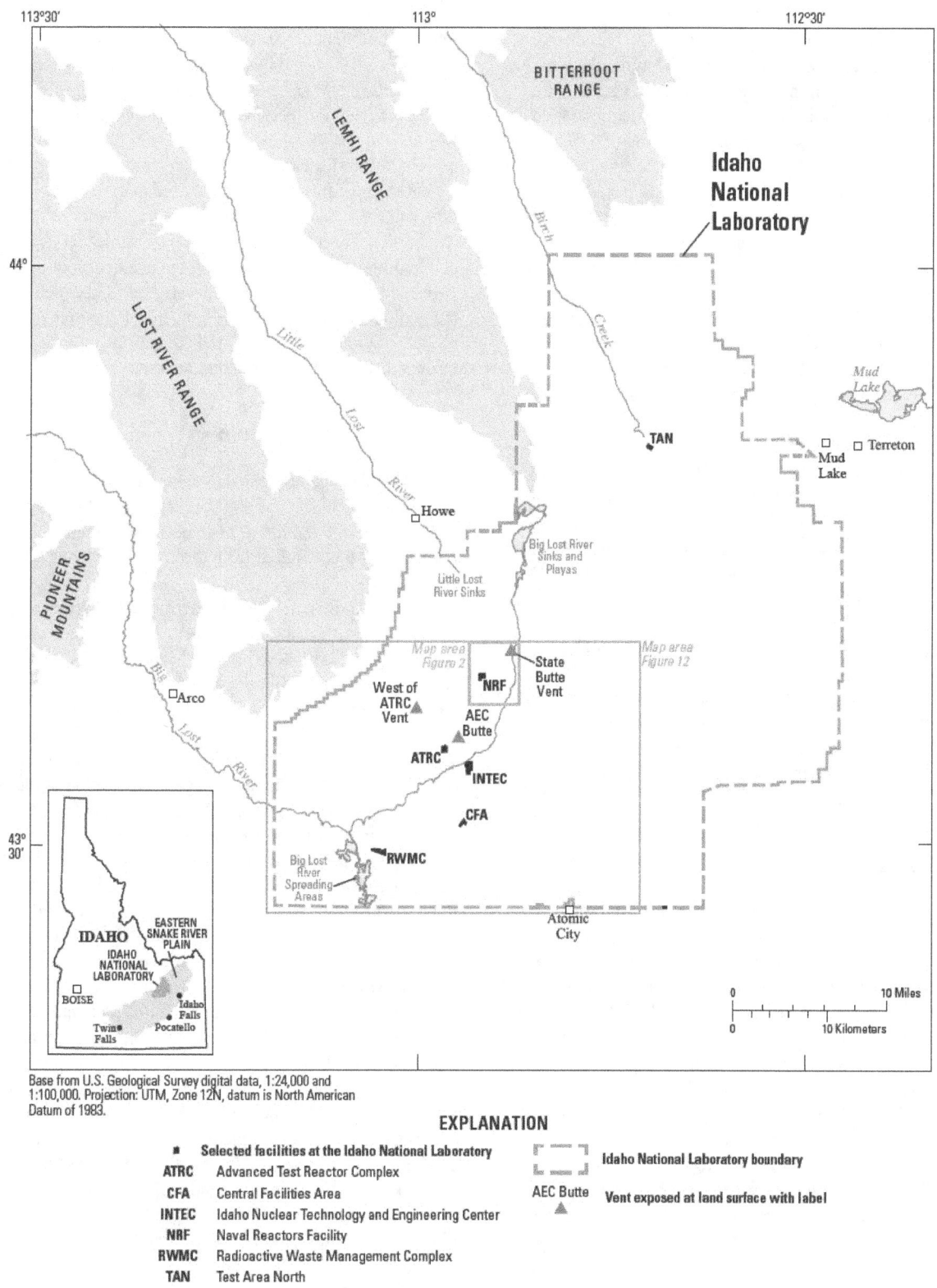

Base from U.S. Geological Survey digital data, 1:24,000 and
1:100,000. Projection: UTM, Zone 12N, datum is North American
Datum of 1983.

EXPLANATION

■	Selected facilities at the Idaho National Laboratory
ATRC	Advanced Test Reactor Complex
CFA	Central Facilities Area
INTEC	Idaho Nuclear Technology and Engineering Center
NRF	Naval Reactors Facility
RWMC	Radioactive Waste Management Complex
TAN	Test Area North

Idaho National Laboratory boundary

AEC Butte Vent exposed at land surface with label
▲

Figure 1.   Location of the eastern Snake River Plain, the boundary of the Idaho National Laboratory (INL), the location of the Naval Reactors Facility (NRF), and other selected facilities.

Paleomagnetic data records the Earth's magnetic field at the time of eruption and is not unique. The remanent magnetization recorded by ferromagnetic minerals in basalt lava flows aligns with the geomagnetic field vector as the basalt crystallizes and cools. The local geomagnetic field vector varies with an average angular motion of 4 to 5 degrees per 100 years in latest Pleistocene and Holocene basalt flows, with extreme variance from 0 to 10 degrees per 100 years (Champion and Shoemaker, 1977). The ESRP monogenetic volcanic fields with volumes of 4 km^3 or more record a single direction of remanent magnetization, which indicates that individual basalt flows that belong to that basalt flow group erupted in a sufficiently brief period of time (less than 100 years) such that any change in the local geomagnetic field vector was too small to detect. Each volcano most likely erupted for only a few days to a few decades.

Paleomagnetic inclination was used to correlate subsurface basalt flow groups based on similar inclination values and polarity. The subcore plugs taken from drill cores and used in this study only yield paleomagnetic inclination data because the original declination was not preserved in the drill cores during drilling. Other data, such as lithology, petrology, geophysical logs, and geochemistry, can be used in conjunction with paleomagnetic inclination and age data to confirm or reject correlations.

Age dates for this study were obtained through conventional potassium-argon (K-Ar) or argon-40/argon-39 (^{40}Ar/^{39}Ar) age experiments. Ten samples from four coreholes were chosen for analysis. Four were dated by using both the K-Ar and the ^{40}Ar/^{39}Ar methods, five were analyzed solely by the K-Ar method, and one was analyzed solely by the ^{40}Ar/^{39}Ar method.

Purpose and Scope

This report presents paleomagnetic and radiometric age information for basalt flow groups from coreholes drilled at and near the Naval Reactors Facility (NRF) at the INL. Coreholes used in this study were drilled between 1989 and 2009. The objective of this study was to describe subsurface stratigraphy derived from paleomagnetic properties and selected ages of basalt flow groups from eight drill cores (USGS 133, NRF B18-1, NRF 89-05, NRF 89-04, NRF 6P, NRF 7P, NRF 15, and NRF 16) taken along a south-to-north transect from corehole USGS 133 through the NRF to corehole NRF 16 at the INL (fig. 2). Subsurface basalt flow groups were then correlated based on their mean paleomagnetic inclinations (appendix A), ages, and stratigraphic position. K-Ar age experiment results from coreholes NRF 89-04 and NRF 89-05 were previously released by Lanphere and others (1993). K-Ar ages for coreholes NRF 6P and NRF 7P are presented in table 1 and ^{40}Ar/^{39}Ar ages for samples from coreholes NRF 6P and NRF 7P were published by Champion and others (2002). Results and analytical data from all these age experiments are presented here for completeness (tables 1, 2, and appendix B).

Results of this study will be used to extend a three-dimensional subsurface stratigraphic framework of the INL (Champion and others, 2011) at and near the NRF for geologic and hydrologic studies, including the evaluation of potential volcanic hazards to the INL. This study will also help to refine the geologic framework of the conceptual and numerical models of groundwater flow and contaminant transport through the ESRP aquifer at and near the INL.

Previous Investigations

Numerous geologic, paleomagnetic, and stratigraphic investigations on surface and subsurface basalts at and near the INL and the ESRP have been conducted. Selected publications that resulted from these investigations are summarized in table 3.

Kuntz and others (1980) described the petrology, subsurface stratigraphic framework, and potential volcanic hazards of the Radioactive Waste Management Complex (RWMC) by using drill core data, which include K-Ar ages, paleomagnetic inclination and polarity data, and petrography. Anderson and Lewis (1989) expanded the stratigraphic framework of the unsaturated zone at the RWMC to numerous other boreholes by using natural gamma geophysical logs as a primary correlation tool in conjunction with core described by Kuntz and others (1980). Anderson (1991) and Anderson and Bowers (1995) used natural gamma logs and sparse core data to describe the stratigraphy of the unsaturated zone and upper part of the ESRP aquifer near the Idaho Chemical Processing Plant [ICPP—now the Idaho Nuclear Technology and Engineering Center (INTEC)], Test Reactor Area [TRA, now the Advanced Test Reactor Complex, (ATRC)] and Test Area North (TAN) areas of the INL. Anderson and others (1996a), Anderson and Liszewski (1997), and Champion and others (2011) summarized the regional stratigraphic relationships of the unsaturated zone and the ESRP aquifer at and near the INL. The Champion and others (2011) report represents the most quantitative interpretation of the regional stratigraphy of the south INL area to date because it is based on paleomagnetic inclination and polarity measurements on cores not previously published, rather than interpretations made from natural gamma geophysical logs.

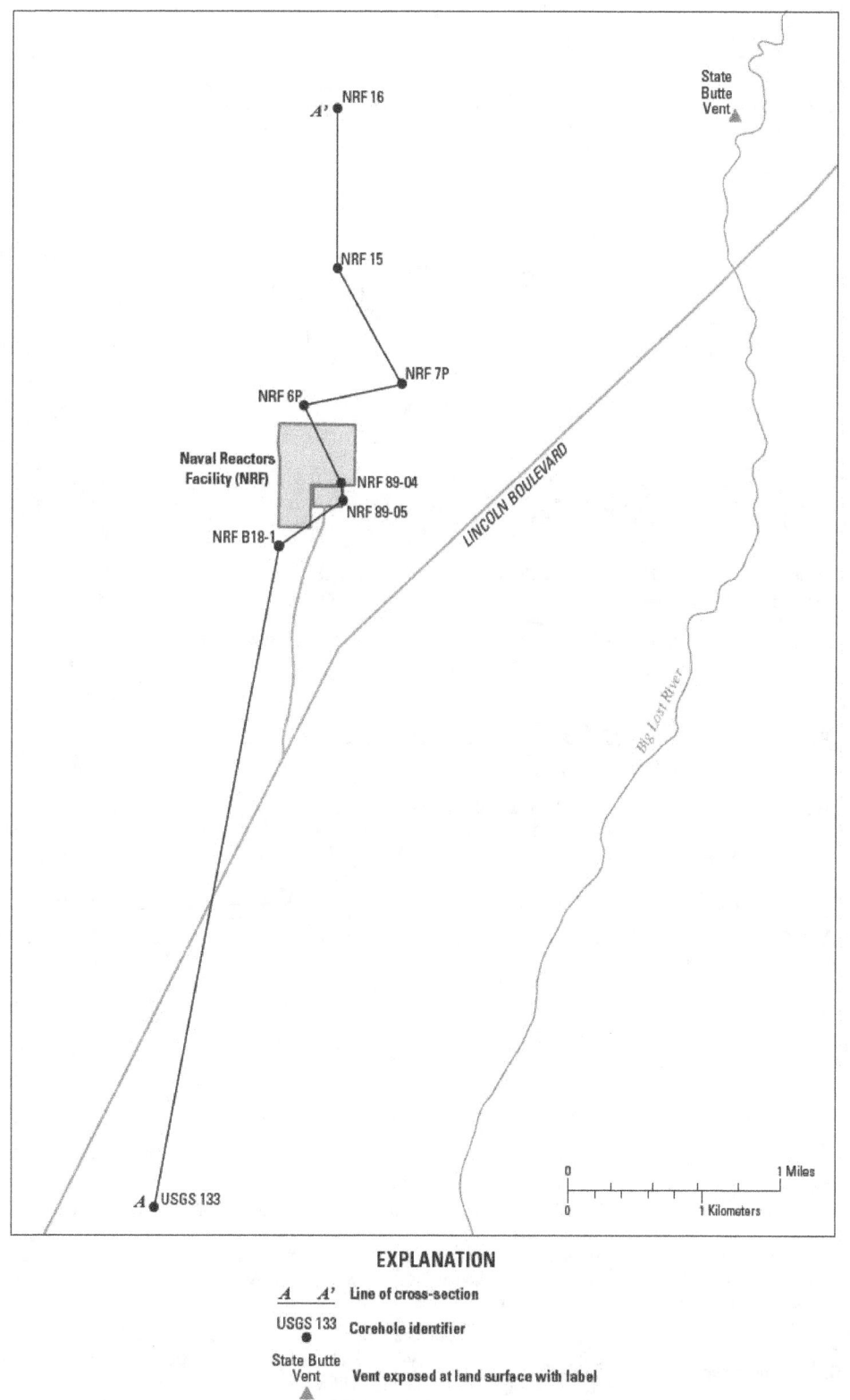

Figure 2.   Location of the Naval Reactors Facility (NRF) at the Idaho National Laboratory, Idaho, the location of eight coreholes described in this report, and the line of cross-section through the eight coreholes.

Table 1.   Weighted mean potassium-argon ages and analytical data for basalt samples from coreholes NRF 89-05, NRF 89-04, NRF 6P, and NRF 7P at the Naval Reactors Facility, Idaho National Laboratory, Idaho.

[Shaded bands indicate results for samples from different cores or depths within the same basalt flow group. **Basalt flow group:** The flow group from which the core sample was collected. ATRC, Advanced Test Reactor Complex; CFA, Central Facilities Area; AEC, Atomic Energy Commission. **Depth (m/ft):** Sample depth is in meters and feet (m/ft). **K$_2$O wt%:** Mean and standard deviation of four measurements in weight percent of potassium oxide measured in each sample. ^{40}Ar$_{rad}$ (10^{-13} mol/g): absolute number of moles/gram (mol/g) of ^{40}Ar measured in each extraction. ^{40}Ar$_{rad}$ (percent): percentage of radiogenic ^{40}Ar measured in each extraction. **Weighted mean flow group age (10^3 years):** weighted mean ages in *red italics* represent the age deemed reasonable for a particular flow group, see text and table 2 for explanations; 10^3 years, 1000s of years]

Core name	Basalt flow group	Experiment number	Depth (m/ft)	K$_2$O wt%	^{40}Ar$_{rad}$ (10^{-13} mol/g)	^{40}Ar$_{rad}$ (percent)	Calculated extraction age (10^3 years)	Weighted mean[1,2] flow group age (10^3 years)
NRF 89-05[3]	West of ATRC	89-05_79a	24/79	0.522±0.002	2.1	2.7	279 ± 45	*303 ± 30*
		89-05_79b			2.435	3.6	323 ± 41	
NRF 89-05[3]	ATRC Unknown Vent	89-05_179a	55/179	0.332±0.001	3.204	4.3	669 ± 75	712 ± 53
		89-05_179b			3.61	4.7	754 ± 74	
NRF 6P	ATRC Unknown Vent	6P_155a	47/155	0.398±0.004	3.483	7.8	609 ± 52	491± 30
		6P_155b			2.395	6.4	418 ± 50	
		6P_155c			2.585	2.6	451 ± 53	
NRF 89-04[3]	CFA Buried Vent	89-04_212a	65/212	0.448±0.002	3.82	1.6	591 ± 149	*492 ± 56*
		89-04_212b			3.067	3.1	475 ± 61	
NRF 89-05[3]	CFA Buried Vent	89-05_235a	72/235	0.646±0.003	5.138	4	552 ± 55	
		89-05_235b			7.047	5.1	758 ± 58	*598 ± 31*
		89-05_235c			4.815	4	518 ± 50	
NRF 89-04[3]	State Butte	89-04_231a	70/231	0.576±0.001	7.161	6	864 ± 58	819 ± 39
		89-04_231b			6.478	6.2	782 ± 53	
NRF 6P	AEC Butte	6P_372a	113/372	0.526±0.001	3.946	1.9	521 ± 110	
		6P_372b			2.597	1.1	343 ± 119	384± 65
		6P_372c			2.178	1.1	287 ± 108	
NRF 7P	North Late Matuyama	7P_440	134/440	0.456±0.003	4.475	2.2	682 ± 121	[4]604 ± 89
		7P_486	148/486	0.463±0.006	3.415	1.6	512 ± 131	

[1]Weighted mean of the calculated extraction ages, where weighting is by the inverse of the variance, and weighted standard deviation.

[2]λε = 0.581 × 10^{-10}yr-1, λB = 4 962 × 10^{-10}yr^{-1}, ^{40}K/K = 1.167 × 10^{-4} mol/mol. Errors are estimates of the standard deviation of analytical precision.

[3]Data from Lanphere and others (1993).

[4]Weighted mean age of two samples from different depths in same flow.

Table 2.   Summary of ^{40}Ar/^{39}Ar ages and analytical data for basalt samples from coreholes NRF 6P and NRF 7P, Naval Reactors Facility, Idaho National Laboratory, Idaho.

[**Basalt flow group:** The flow group from which the core sample was collected. ATRC, Advanced Test Reactor Complex; AEC, Atomic Energy Commission. **Depth (m/ft):** sample depth, in meters and feet (m/ft). **Plateau ^{39}Ar* (percent [steps]):** corrected percentage of ^{39}Ar released (100 percent total) during [n of m] diffusion steps. **Plateau age:** weighted mean age using [n of m] diffusion steps. Plateau ages in *red italics* represent the age deemed reasonable for a particular flow group, see text and table 1 for explanations. **Spectrum description:** qualitative description of the flatness and continuity of the diffusion steps. **Isochron age:** derived from the slope of the best fit line on an isochron diagram plotting ^{40}Ar/^{36}Ar versus ^{39}Ar/^{36}Ar isotopic ratios. **Isochron intercept:** ^{40}Ar/^{36}Ar axis ratio of the best fit isochron line on the isochron diagram. **Abbreviation:** 10^3 years, 1000s of years]

Core name	Basalt flow group	Depth (m/ft)	Plateau ^{39}Ar* (percent [steps])	Plateau age (10^3 years)	Spectrum description	Isochron age (10^3 years)	Isochron intercept
NRF 6P	ATRC Unknown Vent	47/155	98[7 of 8]	*395±25*	Good plateau	372±38	296.6±1.7
NRF 6P	State Butte	74/241	90[5 of 8]	*546±47*	Good plateau	484±89	297.0±1.8
NRF 6P	AEC Butte	113/372	89[4 of 10]	*727±31*	Good plateau	752±63	294.9±1.6
NRF 7P	North Late Matuyama	134/440	50[6 of 10]	*884±53*	Fair plateau	878±242	295.7±3.4
NRF 7P	North Late Matuyama	148/486	94[12 of 18]	1,176 ± 27	Fair plateau	1,178±168	295.5±2.8

Table 3. Summary of selected previous investigations on geology, paleomagnetism, and stratigraphy of the eastern Snake River Plain and Idaho National Laboratory, Idaho.

[**Abbreviations:** INL, Idaho National Laboratory; RWMC, Radioactive Waste Management Complex; INTEC, Idaho Nuclear Technology and Engineering Center (also known as ICPP [Idaho Chemical Processing Plant]); TRA, Test Reactor Area; TAN, Test Area North; ESRP, eastern Snake River Plain; Ma, million years; NPR, New Production Reactor; CFA, Central Facilities Area; NRF, Naval Reactors Facility; USGS; U.S. Geological Survey]

Reference	Area of investigation	Reference summary
Anderson and Lewis, 1989	INL, RWMC	Stratigraphy of the unsaturated zone and upper aquifer at the RWMC, based on natural gamma logs and selected core data.
Anderson, 1991	INL, ICPP and TRA	Stratigraphy of the unsaturated zone and upper aquifer at ICPP and TRA, based on natural gamma logs and selected core data.
Anderson and Bowers, 1995	INL, TAN	Stratigraphy of the unsaturated zone and upper aquifer at ICPP and TRA based on natural gamma logs and selected core data.
Anderson and Liszewski, 1997	INL	INL unsaturated and ESRP aquifer stratigraphy, based on natural gamma logs and selected core data.
Anderson and Bartholomay, 1995	INL, RWMC	RWMC stratigraphy based on natural gamma logs and selected core data.
Anderson and others, 1996a	INL	Stratigraphic data for wells at and near the INL.
Anderson and others, 1999	ESRP, INL, and vicinity	Geologic controls on hydraulic conductivity.
Champion and others, 1981	INL, RWMC	Potassium-argon dating and paleomagnetic data from RWMC basalt cores indicate a 0.46 Ma magnetic reversal.
Champion and others, 1988	INL	Radiometric ages and paleomagnetism at corehole Site E (NPR Test), description of Big Lost cryptochron.
Champion and Herman, 2003	INL, INTEC	Paleomagnetism of basalt from drill cores.
Champion and others, 2002	ESRP, INL	Accumulation and subsidence based on paleomagnetism and selected core data.
Champion and others, 2011	INL	Paleomagnetic correlation of surface and subsurface basalt from drill cores.
Grimm-Chadwick, 2004	INL, CFA	Stratigraphy, geochemistry, and descriptions of high K_2O flow in cores.
Hackett and Smith, 1992	ESRP	Description of ESRP volcanism including development of Axial Volcanic Zone.
Kuntz, 1978	INL, RWMC	Geology of RWMC area.
Kuntz and others, 1980	INL, RWMC	Radiometric dating, paleomagnetism on cores from RWMC.
Kuntz and others, 1986	ESRP	Radiocarbon dates on Pleistocene and Holocene basalt flows.
Kuntz and others, 1992	ESRP	ESRP basaltic volcanism, including eruption styles, landforms, petrology, and geochemistry.
Kuntz and others, 1994	INL	Geologic map of INL, including radiometric ages and paleomagnetism.
Lanphere and others, 1994	INL, TAN	Petrography, age and paleomagnetism of basalt flows at and near TAN.
Lanphere and others, 1993	INL, NRF	Petrography, age and paleomagnetism of basalt flows at and near NRF.
Mazurek, 2004	INL	Genetic alteration of basalt in SRP aquifer.
Miller, 2007	INL, RWMC	Geochemistry, and descriptions of the B flow, and stratigraphy of corehole USGS 132.
Morse and McCurry, 2002	INL	Base of the aquifer, alteration in basalts.
Reed and others, 1997	INL, ICPP	Geochemistry of lava flows in cores at ICPP.
Rightmire and Lewis, 1987	INL, RWMC	Unsaturated zone geology, geochemistry of sediment and alteration.
Scarberry, 2003	INL, RWMC	Geochemistry of the F flow (now referred to as the Big Lost Reversed Polarity Cryptochron flows) and distribution in several coreholes at the INL.
Shervais and others, 2006	INL, TAN	Cyclic geochemical variations in basalt in TAN drill cores.
Tauxe and others, 2004	SRP	Paleomagnetism of the Snake River Plain.
Twining and others, 2008	INL	Construction diagrams, lithological, and geophysical logs for coreholes.
Welhan and others, 2002	INL, ESRP	Morphology of inflated pahoehoe flows.
Wetmore and Hughes, 1997	INL	Model morphologies of subsurface lava flows.
Wetmore and others, 1999	INL	Axial Volcanic Zone construction.

Basalt Flow and Flow Group Definition and Labeling Conventions

In this report, a basalt flow is defined as a single, continuous outpouring of basalt on the Earth's surface. A basalt flow group is defined as any and all basalt flows from a single monogenetic volcanic eruption. The ESRP monogenetic volcanic fields record a single direction of paleomagnetic inclination, which indicates that individual basalt flows that belong to that flow group were emplaced during a few days to a few decades of time (Kuntz and others, 1992). In this report, basalt flow groups were identified by boundaries where polarity or mean inclination values changed. Some boundaries were identified by sedimentary interbeds, oxidized flow tops, or areas where a significant change in petrography was evident.

Whenever possible, basalt flow groups were labeled for their vents, such as "AEC Butte flow group," and are so labeled on cross-sections. Flow groups without known surface vents were labeled for chemical, spatial, or paleomagnetic identifiers. For example, the Jaramillo (Matuyama) flow group is named for the Jaramillo Normal Polarity Subchron of the Matuyama Reversed Polarity Chron (fig. 3). Other names were derived from proximity to INL facilities, such as the ATRC, and may have designations such as West of ATRC flow group. For convenience, the labeled basalt flow groups are referred to as a "flow group," rather than a "basalt flow group."

Geologic Setting

The INL covers an area of about 2,300 km^2 on the ESRP, which is a northeast-trending structural trough that contains Neogene volcanic rocks and interbedded sediments (fig. 1). Tertiary rhyolites are generally exposed only along the margins of the ESRP, but several Quaternary rhyolite domes surrounded by basalt flows are located on the ESRP. Late Pleistocene and Holocene basalt fields cover most of the ESRP. The axis of the ESRP is a locus of the basalt shield vents from which young basalt flows erupted. Kuntz and others (1994) showed that the surface area of the INL is dominated by basalt flows of the Qbb (basaltic lava flows and pyroclastic deposits, upper to middle Pleistocene, estimated to be 15-200 ka) and Qbc (basaltic lava flows and pyroclastic deposits, middle Pleistocene, estimated to be 200-400 ka) relative age classification units, which suggests that they are 15 to 400 thousand years (ka) in age.

Basalt flows at the INL are very similar petrographically. All are tholeiitic olivine basalt that contains olivine, plagioclase, clinopyroxene, ilmenite, magnetite, glass, and accessory apatite. Basalt flows may have significantly different textures despite having similar mineralogy and chemistry. In the subsurface, sedimentary interbeds separate some basalt flows; the distribution and thickness of these interbeds varies. Sedimentary interbeds are important in the subsurface correlation at the NRF.

Basalt flow groups are composed of one or more individual basalt flows that range from 1.5 to about 28 m thick. Sequences of several thin flows may indicate proximity to the vent for that basalt flow group. A basalt flow of greater than about 9 m may indicate a ponded or inflated flow that accumulated in a depression where the outlet was restricted or blocked. These depressions may have formed in strictly geomorphic ways, such as a closed basin between several shield volcanoes, or a depression along the course of an ephemeral stream. They may have formed due to topography created by earlier basalt flows of the same eruption of which they are part. The depressions may also have formed in structural ways, such as slumping of surface blocks in volcanic rift zones, slumping related to compaction of underlying sedimentary or volcanic deposits, or perhaps faulting.

Most flows investigated in this study do not have significant amounts of cinder or rubble at or near their uppermost surfaces. This suggests that most of the flows are pahoehoe and not slab pahoehoe or aa flows in morphology.

Surface basalt flows and vents in the south and central parts of the INL have normal magnetic polarity (Kuntz and others, 2007) and were erupted during the Brunhes Normal Polarity Chron (less than 0.78 Ma) (fig. 3; Ogg and Smith, 2004). Some surface basalt flows and vents located just to the north and west of the NRF have reversed polarity (Lanphere and others, 1994) and were erupted during the Matuyama Reversed Polarity Chron (2.58–0.78 Ma) (fig. 3; Ogg and Smith, 2004). A reversed polarity basalt flow group is sometimes found at depths of from 107 to 152 m in coreholes from the southern INL area and was identified as the Big Lost Polarity Cryptochron basalt flow group as shown in figure 3 and by Champion and others (1988). This basalt flow group was not found in any of the coreholes described from the NRF area.

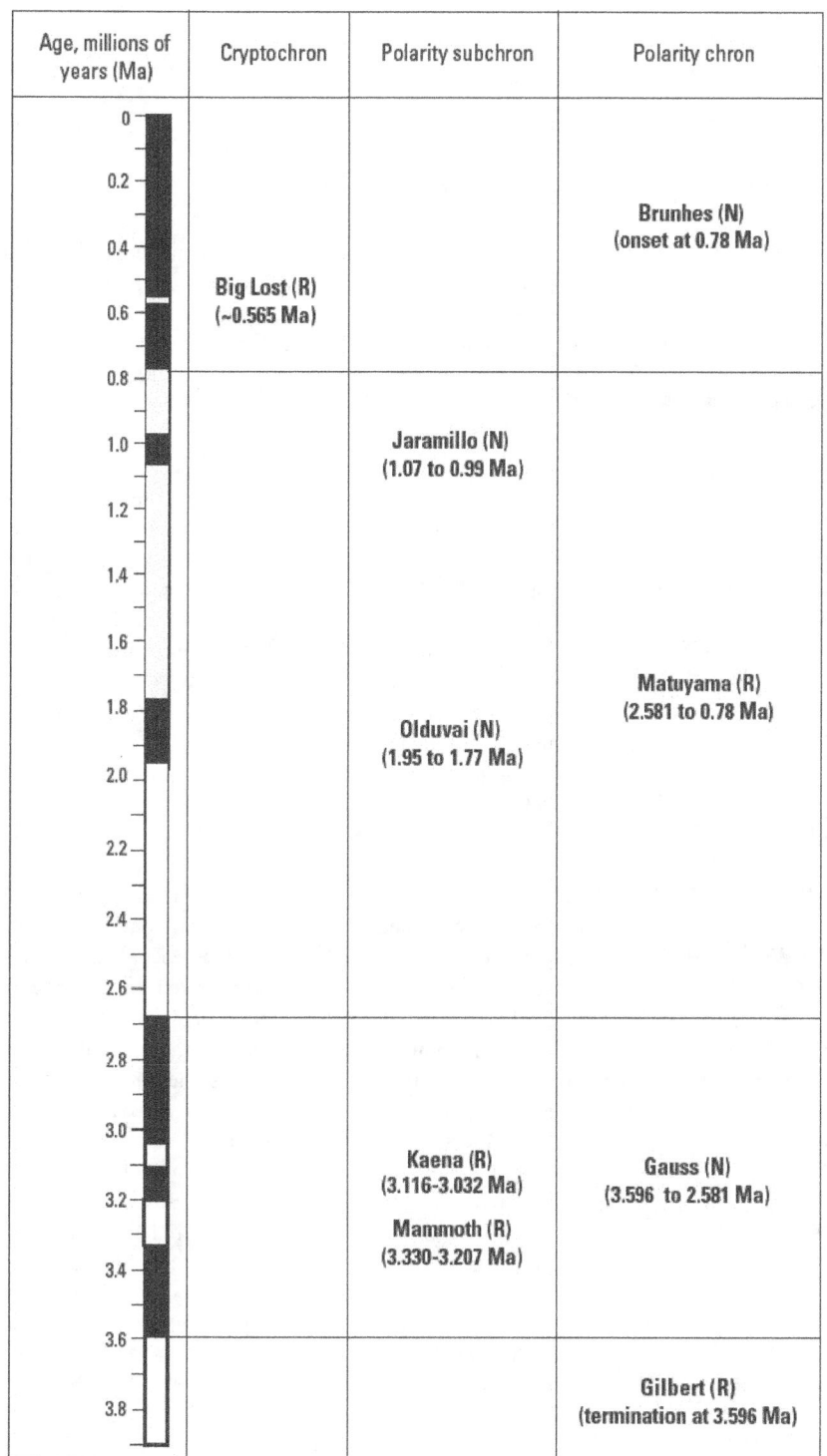

Age, millions of years (Ma)	Cryptochron	Polarity subchron	Polarity chron
	Big Lost (R) (~0.565 Ma)		Brunhes (N) (onset at 0.78 Ma)
		Jaramillo (N) (1.07 to 0.99 Ma)	
		Olduvai (N) (1.95 to 1.77 Ma)	Matuyama (R) (2.581 to 0.78 Ma)
		Kaena (R) (3.116-3.032 Ma) Mammoth (R) (3.330-3.207 Ma)	Gauss (N) (3.596 to 2.581 Ma)
			Gilbert (R) (termination at 3.596 Ma)

EXPLANATION

■ Black bars in age column indicate normal polarity (N - normal polarity)

☐ White bars in age column indicate reversed polarity (R - reversed polarity)

Figure 3. Pliocene-Holocene geomagnetic time scale.

Analytical Methods

The drill cores from coreholes USGS 133, NRF B18-1, NRF 89-04, NRF 89-05, NRF 6P, NRF 7P, NRF 15, and NRF 16, were carefully logged and sampled by using INL Lithologic Core Storage Library protocols as described in Davis and others (1997). Prior to sampling, the core material was described, and the tops and bottoms of basalt flows were identified. Depths were measured by tape measure from known marks recorded on foam markers or depth markings on the cores in the core boxes. The foam markers or depth markings were placed by the drillers at the time of coring and recorded the measured depths logged at the end of each coring run. Detailed petrographic descriptions were not made for the cores in this study.

For paleomagnetic studies, an attempt was made to take seven samples of core from each basalt flow identified. For each sample, a 2.5-cm diameter subcore was drilled at right angles to the axis of the original core to provide material for paleomagnetic analysis. The samples were trimmed to 2.2-cm lengths, and the inclination and intensity of magnetization were measured with a cryogenic magnetometer. Progressive alternating-field demagnetization was performed on nearly all samples to remove any components of secondary magnetization. A few samples from coreholes NRF 6P, NRF 15, and NRF 16, particularly near polarity transitions, were thermally demagnetized to better remove opposite polarity that overprinted magnetizations. The primary or characteristic remanent inclinations measured on individual basalt samples are shown in figures 4 through 11. Mean inclination values for each basalt flow group and 95-percent confidence limits about the mean value were calculated by using the method of McFadden and Reid (1982) and are listed in appendix A. The subcore plugs taken from drill cores and used in this study only yield inclination data because the original declination was not preserved in the drill cores during drilling.

To facilitate paleomagnetic interpretations, the coreholes were assumed to be vertical in their original drilling orientation. Gyroscopic deviation logs of coreholes NRF 15 and NRF 16 were made at 0.3-m intervals, and they record moderate deviations to the east-northeast, no greater than 2.0 and 2.4 degrees from vertical, respectively, for those coreholes. A gyroscopic deviation log also was made for corehole USGS 133 with data recorded at 1.5-m intervals and recorded deviation no greater than 1.5 degrees from vertical to the northwest. Because these small deviations are not within the north-to-south vertical plane, no correction needed to be applied to the measured paleomagnetic remanent inclination values for samples from these coreholes. No gyroscopic logs were available for coreholes NRF B18-1, NRF 89-05, NRF 89-04, NRF 6P, and NRF 7P. Deviation from vertical for any particular depth interval in INL coreholes typically is less than one degree and does not significantly affect paleomagnetic remanent inclination interpretations.

Paleomagnetic Results

Paleomagnetic measurements were made on 749 basalt samples from the eight coreholes. The inclinations for the individual basalt flows were used to determine the mean for the basalt flow group. The mean inclination values for individual basalt flow groups ranged from -40.3 to 77.0 degrees within the typical range of inclinations owing to geomagnetic secular variation (Champion and Shoemaker, 1977). The precision of these mean inclination values, which were reported as two standard errors of the mean (α95), ranges between 0.6 and 17.0 degrees, while the mean, median, and mode of this two standard error population are identical at 1.4 degrees. Data of this precision allow the differentiation and correlation of basalt flows or basalt flow groups combined with the stratigraphic constraints from the physical core logging and qualitative petrographic assessments.

Basalt flow group assignments were made according to mean paleomagnetic inclination and polarity data and stratigraphic position. Similar mean inclination values, however, can be randomly acquired by two successive basalt flows or flow groups, even though they may be tens of thousands of years different in age. Flows with different petrography that have the same mean inclination values and that are not separated by a sedimentary interbed may have erupted from two different vents and represent two different magma batches. This phenomenon has been demonstrated in the surface paleomagnetic and petrographic/chemical work presented in Kuntz and others (2002). For the basalt flow groups analyzed in this report, it was sometimes difficult to distinguish between these interpretive alternatives. These petrographic assessments were qualitative and do not replace any future quantitative chemical and petrographic studies of these coreholes. Changes in phenocryst abundances, however, were noted for the different basalt flows in the NRF coreholes and were used to help identify significant boundaries between these flows.

As shown in plate 1, 13 basalt flow groups were identified in the 8 coreholes and range in thickness from 2.4 to 56 m. Of the 13 basalt flow groups identified in the 8 coreholes, 7 were erupted during the Brunhes Normal Polarity Chron (0.78–0 Ma), and 6 were erupted during the latter part of the Matuyama Reversed Polarity Chron (2.58–0.78 Ma), which included 1 that was erupted and magnetized during the Jaramillo Normal Polarity Subchron (1.07–0.99 Ma) of the Matuyama Reversed Polarity Chron (fig. 3; pl. 1; Ogg and Smith, 2004). The results for individual coreholes are discussed below.

Corehole USGS 133

Corehole USGS 133 is a 248-m-deep corehole located about 5 km south-southwest of the NRF (fig. 2; pl. 1). Corehole USGS 133 contains eight basalt flow groups separated by conformable contacts and sedimentary interbeds. A log of the core stratigraphy, paleomagnetic inclinations, and the depths of samples used for paleomagnetic measurements is shown in figure 4. The individual and mean inclination values and uncertainties and depth ranges for the inclination values for corehole USGS 133 are given in appendix A.

Corehole NRF B18-1

Corehole NRF B18-1 is a 77-m-deep corehole located 122 m southwest of the NRF (fig. 2; pl. 1). Corehole NRF B18-1 contains three basalt flow groups separated by conformable contacts and sedimentary interbeds. A log of the core stratigraphy, paleomagnetic inclinations, and the depths of samples used for paleomagnetic measurements is shown in figure 5. The individual and mean inclination values and uncertainties and depth ranges for the values for corehole NRF B18-1 are given in appendix A.

Corehole NRF 89-05

Corehole NRF 89-05, which was completed in September 1989, is a 74-m-deep corehole located near the center of the NRF (fig. 2; pl. 1). Corehole NRF 89-05 contains three basalt flow groups separated by sedimentary interbeds. A log of the core stratigraphy, paleomagnetic inclinations, and the depths of samples used for paleomagnetic measurements is shown in figure 6. The individual and mean inclination values and uncertainties, and depth ranges for the values for corehole NRF 89-05 are shown in appendix A.

Corehole NRF 89-04

Corehole NRF 89-04, which was completed in September 1989, is a 76-m-deep corehole also located near the center of the NRF (fig. 2; pl. 1). Only the deepest part of the core was sampled because of the proximity of this corehole to corehole NRF 89-05, which was sampled in its entirety. The sampled interval of corehole NRF 89-04 ranges from 59.7 to 75.6 m below land surface (BLS) and contains two basalt flow groups separated by a sedimentary interbed. A log of the core stratigraphy, paleomagnetic inclinations, and the depths of samples used for paleomagnetic measurements is shown in figure 7. The individual and mean inclination values and uncertainties and depth ranges for the values for corehole NRF 89-04 are given in appendix A.

Corehole NRF 6P

Corehole NRF 6P is a 152.4-m-deep corehole located about 425 m north-northwest of the center of the NRF (fig. 2; pl. 1). Corehole NRF 6P contains six basalt flow groups separated by conformable contacts and sedimentary interbeds. A log of the core stratigraphy, paleomagnetic inclinations, and the depths of samples used for paleomagnetic measurements is shown in figure 8. The individual and mean inclination values and uncertainties and depth ranges for the values for corehole NRF 6P are given in appendix A.

Corehole NRF 7P

Corehole NRF 7P is a 153-m-deep corehole located about 1,040 m northeast of the center of the NRF (fig. 2; pl. 1). Corehole NRF 7P contains seven basalt flow groups separated by conformable contacts and sedimentary interbeds. A log of the core stratigraphy, paleomagnetic inclinations, and the depth of samples used for paleomagnetic measurements is shown in figure 9. The individual and mean inclination values and uncertainties and depth ranges for the values for corehole NRF 7P are given in appendix A.

Corehole NRF 15

Corehole NRF 15 is a 231-m-deep corehole located 1.5 km north of the center of the NRF (fig. 2; pl. 1). Coring began at a depth of 4.9 m, recovering basalt samples; sediments are present above that depth. Corehole NRF 15 contains 10 basalt flow groups separated by conformable contacts and sedimentary interbeds. A log of the core stratigraphy, paleomagnetic inclinations, and the depths of samples used for paleomagnetic measurements is shown in figure 10. The individual and mean inclination values and uncertainties, and depth ranges for the values for corehole NRF 15 are given in appendix A.

Corehole NRF 16

Corehole NRF 16 is a 129-m-deep corehole located 2.6 km north of the center of the NRF (fig. 2; pl. 1). Corehole NRF 16 contains six basalt flow groups separated by conformable contacts and sedimentary interbeds. A log of the core stratigraphy, paleomagnetic inclinations, and the depths of samples used for paleomagnetic measurements is shown in figure 11. The individual and mean inclination values and uncertainties and depth ranges of the values for corehole NRF 16 are given in appendix A.

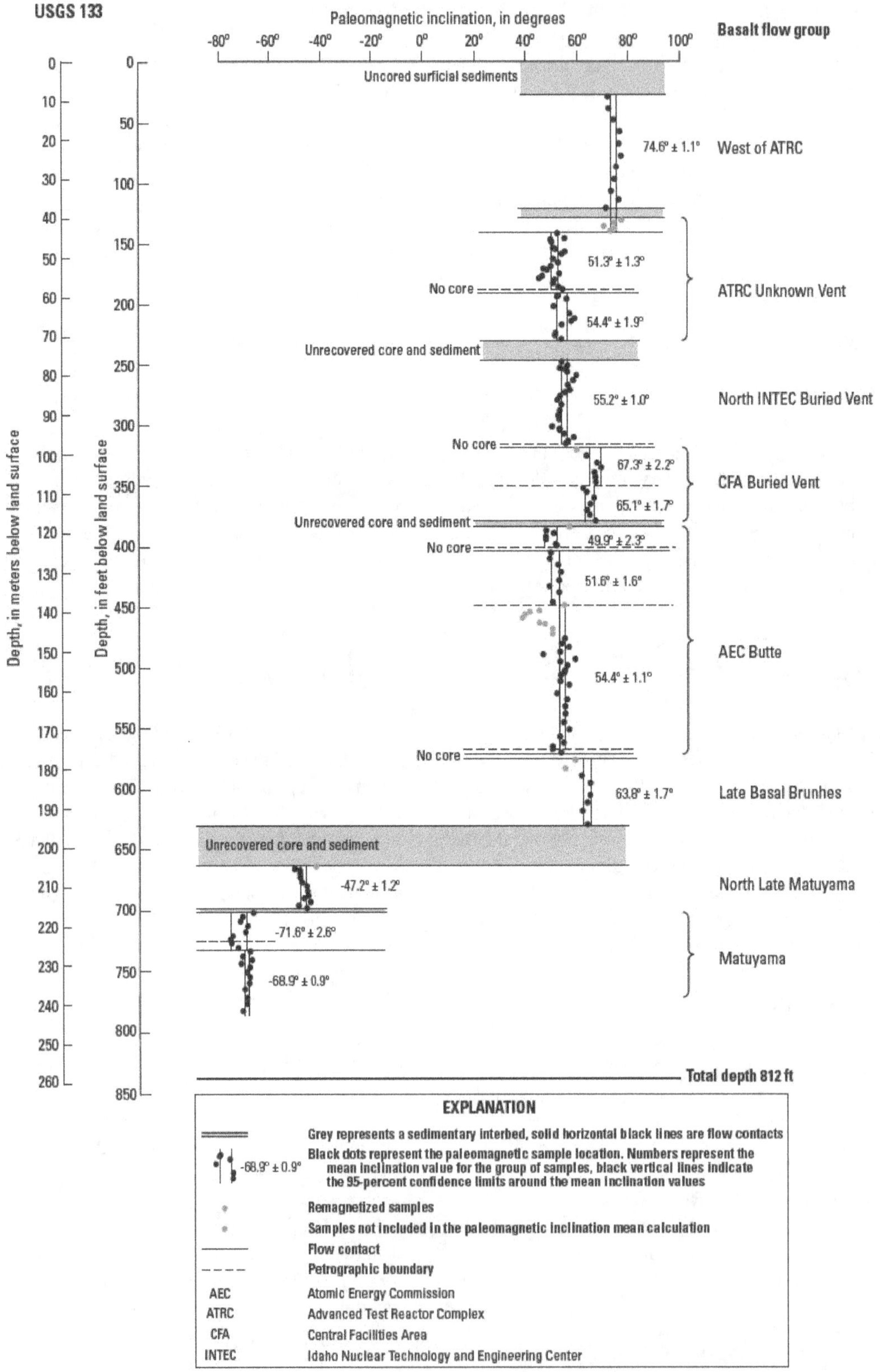

Figure 4.   Log of the core stratigraphy and mean paleomagnetic inclination data for corehole USGS 133 at the Idaho National Laboratory, Idaho.

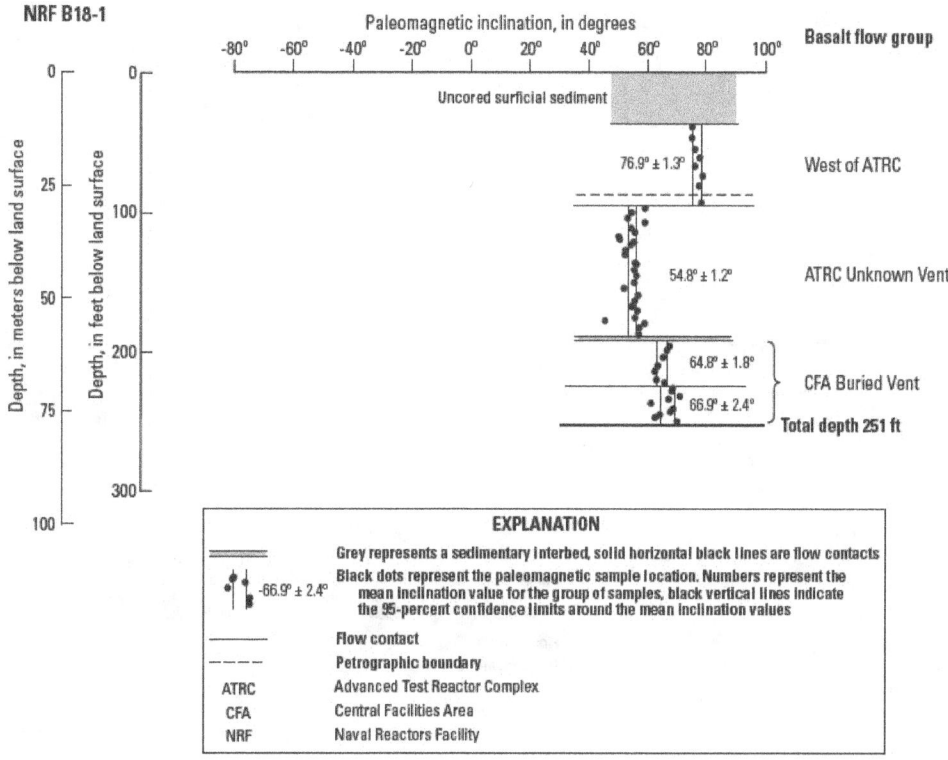

Figure 5. Log of the core stratigraphy and mean paleomagnetic inclination data for corehole NRF B18-1 at the Idaho National Laboratory, Idaho.

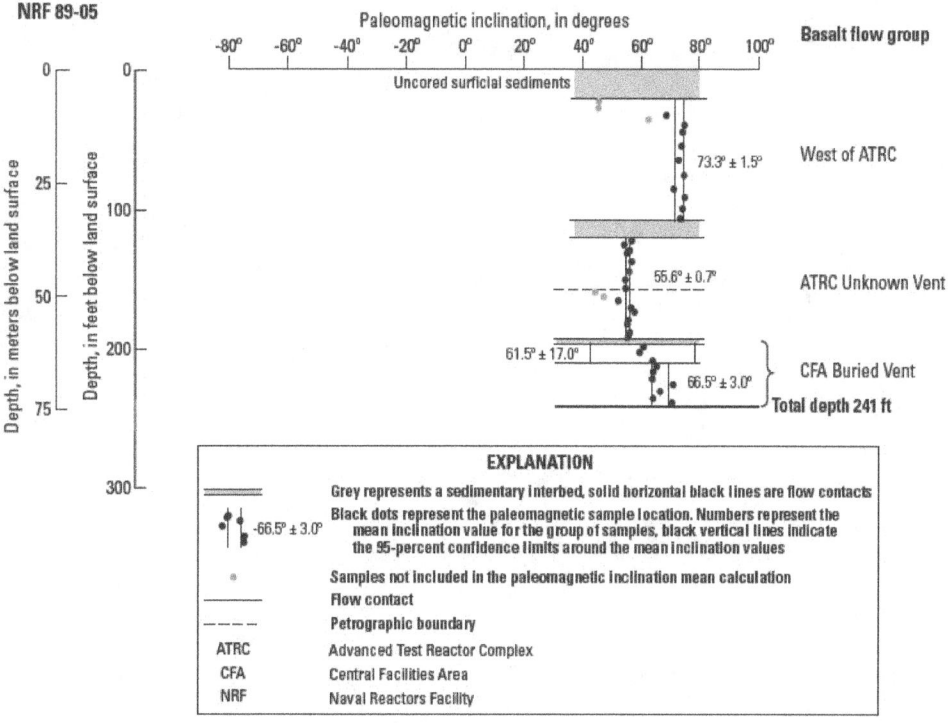

Figure 6. Log of the core stratigraphy and mean paleomagnetic inclination data for corehole NRF 89-05 at the Idaho National Laboratory, Idaho.

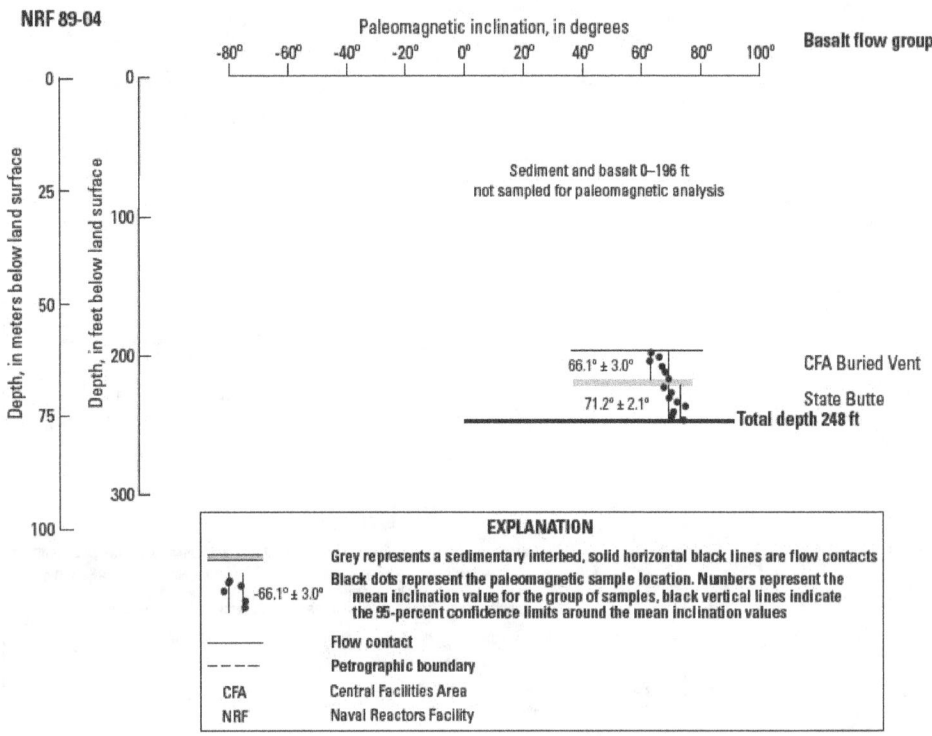

Figure 7.   Log of the core stratigraphy and mean paleomagnetic inclination data for corehole NRF 89-04 at the Idaho National Laboratory, Idaho.

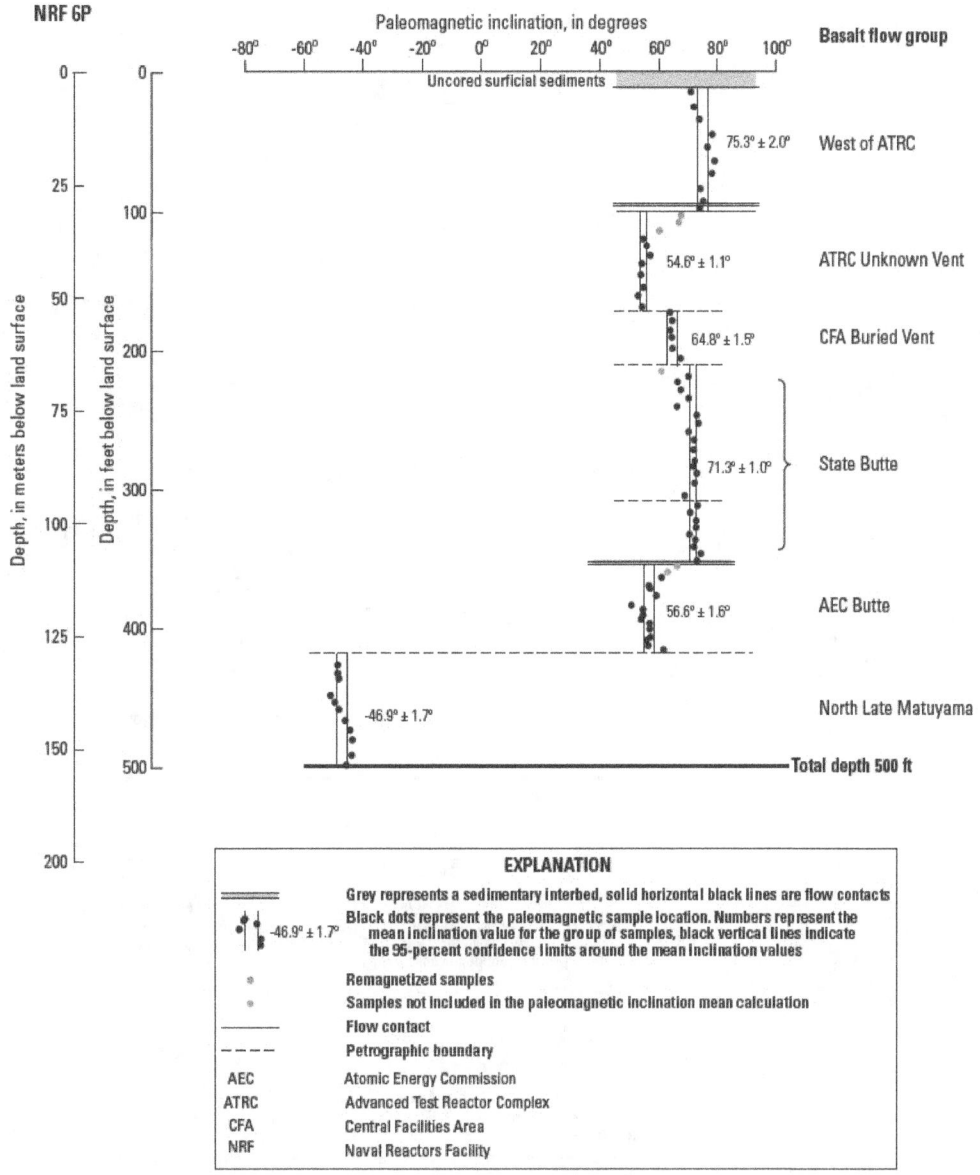

Figure 8.   Log of the core stratigraphy and mean paleomagnetic inclination data for corehole NRF 6P at the Idaho National Laboratory, Idaho.

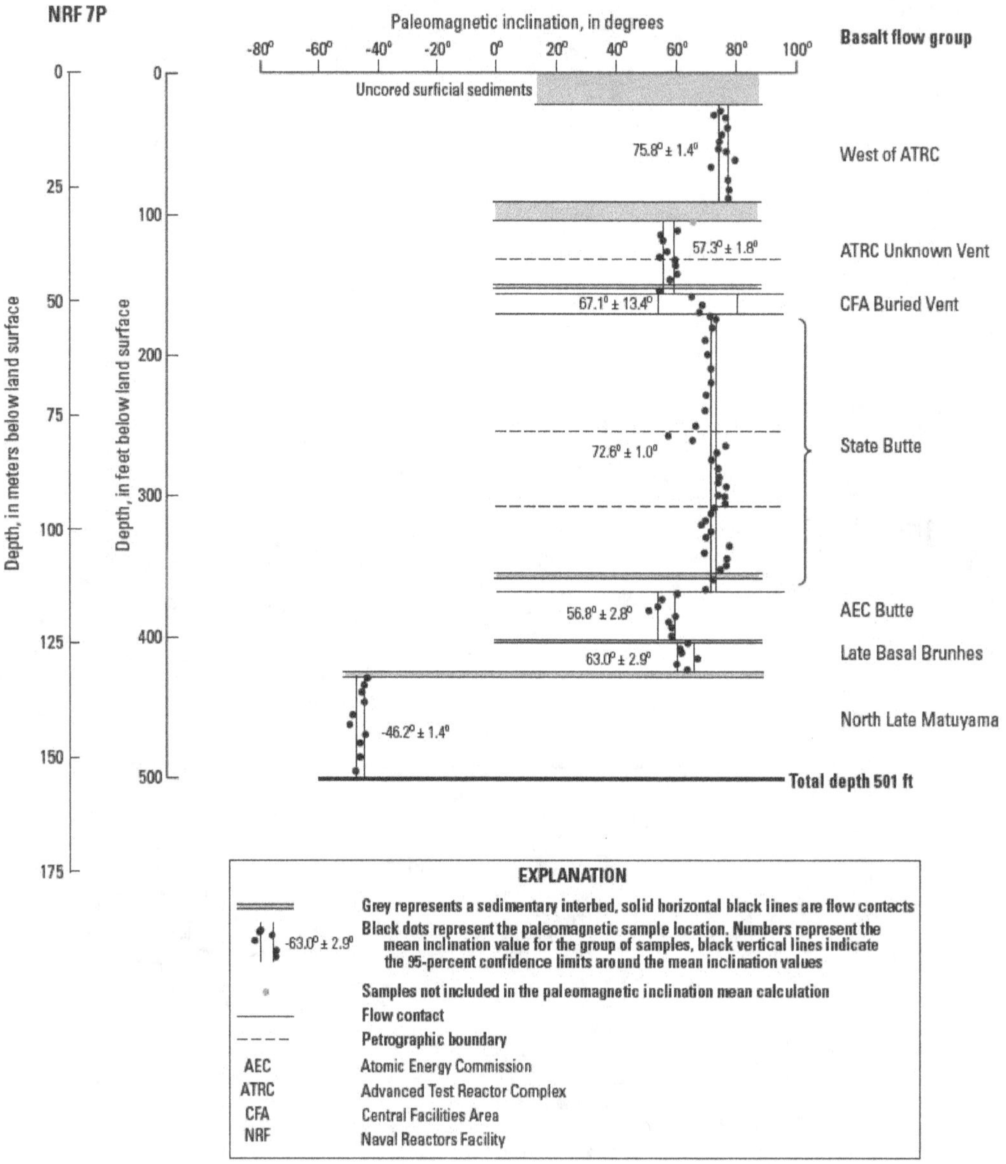

Figure 9.   Log of the core stratigraphy and mean paleomagnetic inclination data for corehole NRF 7P at the Idaho National Laboratory, Idaho.

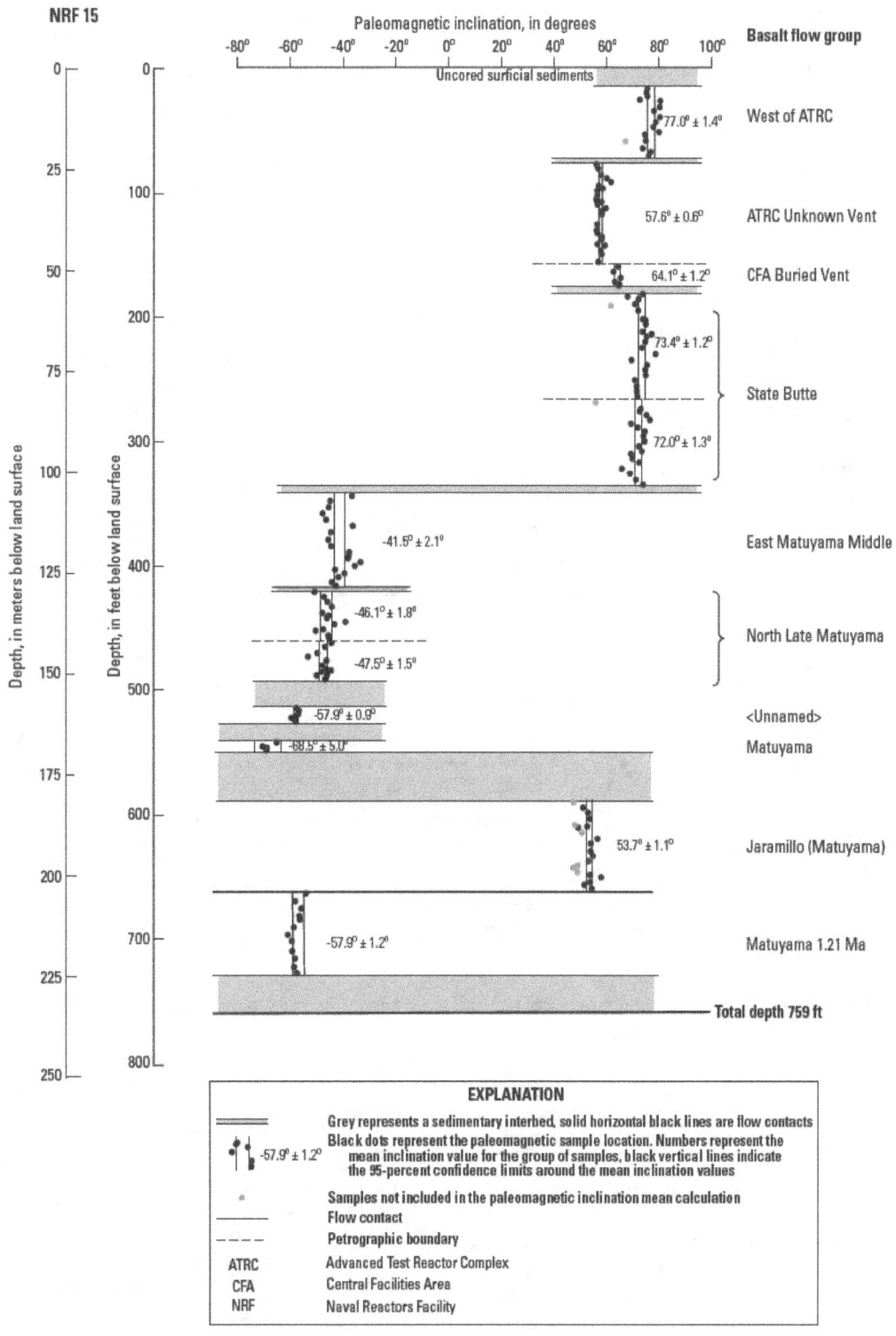

Figure 10. Log of the core stratigraphy and mean paleomagnetic inclination data for corehole NRF 15 at the Idaho National Laboratory, Idaho.

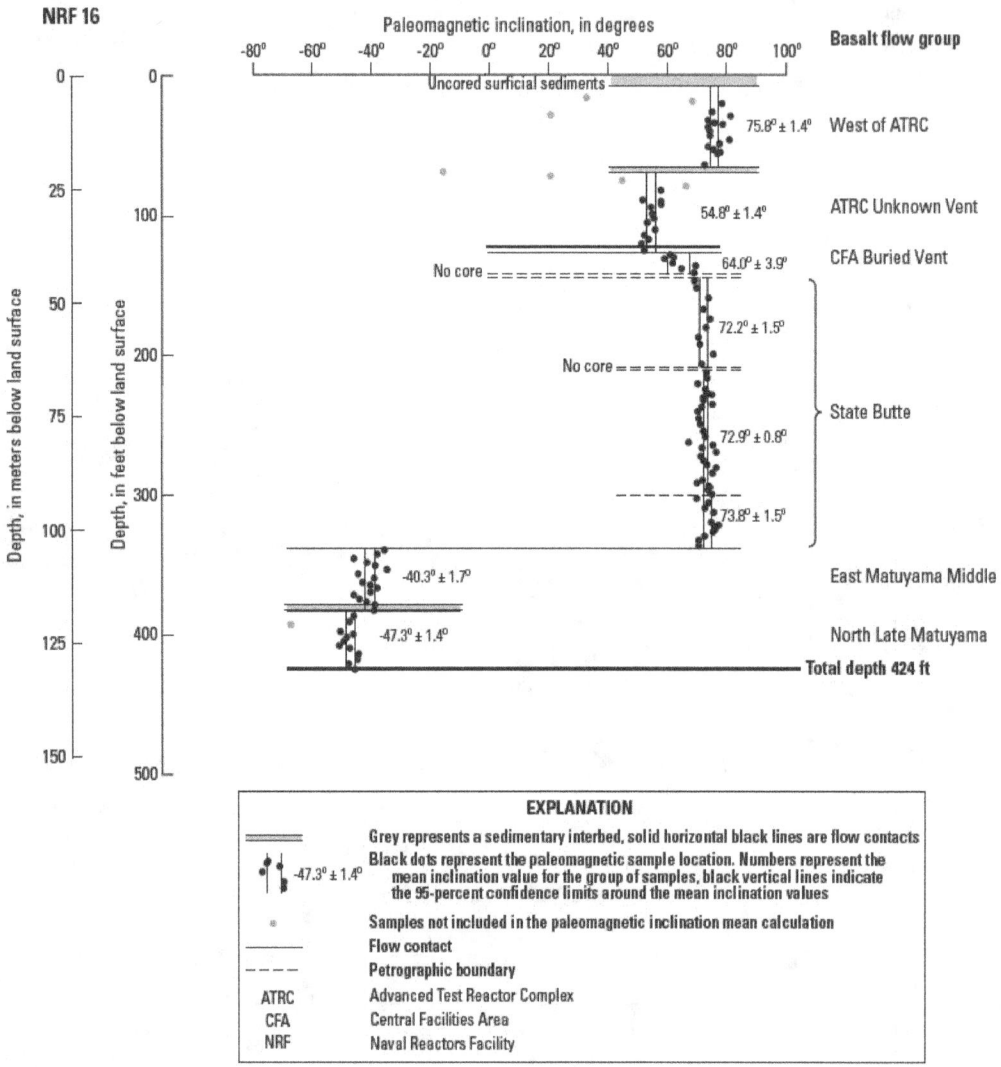

Figure 11.   Log of the core stratigraphy and mean paleomagnetic inclination data for corehole NRF 16 at the Idaho National Laboratory, Idaho.

Age Data

Age experiments that use conventional K-Ar and $^{40}Ar/^{39}Ar$ techniques were obtained from samples from four of the coreholes (NRF 89-05, NRF 89-04, NRF 6P, and NRF 7P) and are presented in this report to help better constrain stratigraphic interpretation. A total of 10 samples from the four coreholes were chosen according to the criteria of acceptability for K-Ar dating (Dalrymple and Lanphere, 1969; Mankinen and Dalrymple, 1972) or $^{40}Ar/^{39}Ar$ age dating. Thin sections of basalt from these coreholes were examined

petrographically to determine those most suitable for age dating. The samples were selected to minimize the amount of glass in the groundmass. In these samples, the volume of glass in the groundmass was considered to be too small to affect the measurements. The samples were crushed and ground to a size of 0.5 to 1 mm. Of these 10 samples, 4 were dated by using K-Ar and $^{40}Ar/^{39}Ar$ methods, 5 were analyzed solely by the K-Ar method, and 1 was analyzed solely by the $^{40}Ar/^{39}Ar$ method (tables 1, 2; appendix B). Results of five of the conventional K-Ar experiments on samples from coreholes NRF 89-04 and NRF 89-05 are shown in table 1 and were previously reported by Lanphere and others (1993).

Limitations of Radiometric Dating of Olivine Tholeiite Basalts

ESRP olivine tholeiite basalts have generally less than 2 weight percent potassium (Stout and Nicholls, 1977, table 9), and the basalts used in this study are young, less than 1 Ma. Age experiments rely on accurate analyses of small quantities of potassium and the daughter products of its radiogenic isotope, which may be difficult in basalts with little potassium. Conventional K-Ar age experiments can be skewed to old ages owing to excess ^{40}Ar embedded in phenocryst phases that grew in the magma chamber because the method uses a single argon gas extraction.

Discrepancies between conventional K-Ar and ^{40}Ar/^{39}Ar age experiments arise from inherent procedural constraints of the two dating methods. In conventional K-Ar experiments, independent chemical and mass spectroscopic analyses are conducted on separate aliquots in precise and complete extractions of the elements being studied. The argon part of the analysis is a single thermal diffusion presumed to be a total extraction. Concentrations of potassium and argon by analysis are combined mathematically, and the age of the sample is calculated. Although conventional K-Ar age experiments can be accurate (Lanphere, 2000), the expectation that a single thermal extraction will yield argon gas of homogeneous isotopic ratios is sometimes violated. The ^{40}Ar/^{39}Ar method avoids using this analytical assumption by performing progressive thermal diffusion extractions to prove that the argon isotopic ratios are homogeneous. Also, the analysis is conducted on a single, small-mass sample after it has been irradiated by fast neutrons to convert some of the potassium to argon.

Therein lies the potential flaw in ^{40}Ar/^{39}Ar dating experiments on young, fine-grained basalt samples. Fast neutron irradiation can cause the resulting ^{39}Ar, ^{37}Ar, and ^{36}Ar ions to "recoil" out of their crystallographic domain and into adjacent domains that may thermally diffuse at higher or lower temperatures. This causes progressive change and abrupt discontinuities in the argon isotopic ratios being produced in successive temperature steps of extraction. The irradiation timeframe must be held to a minimum, gas samples being extracted and analyzed on the mass spectrometer must be of reasonably uniform size, and active knowledge of the mass discrimination value of the instrument must be maintained. Despite the refinements in technique listed above, some aphanitic basalt samples cannot be dated by using the ^{40}Ar/^{39}Ar method.

The actual age of a basalt flow at the INL may be known only to the nearest 100 thousand years (k.y.) or so despite the analytical uncertainty from any single age experiment. Given the assumptions and uncertainties of both methods, the results for each method must be evaluated and the age date that seems most reasonable based on known stratigraphy and other data must be determined. Numerous age dates derived by both methods are in proper depth order and replication of ages on correlative flows in different coreholes has been shown (Champion and others, 2002, 2011). Repeated age experiments using the conventional K-Ar and ^{40}Ar/^{39}Ar age methods on different samples of the same basalt flow may identify the potential flaws in both approaches and allow a reasonable age to be chosen.

Methods of Analysis

K-Ar Age Experiments

An aliquot (10 g) of the 0.5- to 1-mm master sample was pulverized to less than 74 microns. This powdered material was used for the potassium oxide (K_2O) measurements. The K_2O measurements were made in duplicate on each of two splits of powder by flame photometry after lithium metaborate fusion and dissolution (Ingamells, 1970). Argon analyses were made by isotope dilution mass spectrometry using techniques described by Dalrymple and Lanphere (1969). Aliquots of the 0.5- to 1-mm master sample were baked overnight in a vacuum in an argon extraction system at 280°C. Argon mass analyses were done on a computerized multiple-collector mass spectrometer with a 22.86-cm radius and a nominal 90-degree sector magnet (Stacey and others, 1981). The analytical error for an individual age measurement was calculated by using the method of Cox and Dalrymple (1967). Weighted mean ages for basalt flows were calculated by using the method of Taylor (1982). Results from K-Ar experiments for the four coreholes are shown in table 1.

^{40}Ar/^{39}Ar Age Experiments

Splits of whole-rock samples were packaged in copper foil and placed in cylindrical vials with mineral grains of known age. Samples were irradiated with fast neutrons along with a monitor mineral of known age [sanidine from the Taylor Creek Rhyolite, 27.92 Ma (Dalrymple and Duffield, 1988)], to induce the nuclear reaction ^{39}K (neutron, proton)^{39}Ar. Ages of samples were calculated from the measured ^{40}Ar/^{39}Ar ratio after determining the fraction of ^{39}K converted to ^{39}Ar by analyzing the monitor mineral. Corrections for interfering argon isotopes produced from potassium and calcium and for atmospheric argon were also made.

Results of Age Experiments by Basalt Flow Group

West of Advanced Test Reactor Complex (ATRC) Flow Group

The West of ATRC flow group (pl. 1) was sampled for K-Ar age experiments from corehole NRF 89-05 from 24.1-m BLS. The experiment produced a weighted mean age of 303 ± 30 ka from two separate extractions [Experiment numbers (Exp. nos.) 89-05_79a and 89-05_79b; table 1]. Paleomagnetic inclination correlation of this sample to coreholes south of the NRF indicates that this age seems reasonable (Champion and others, 2011).

Advanced Test Reactor Complex (ATRC) Unknown Vent Flow Group

The ATRC Unknown Vent flow group (pl. 1) was sampled for K-Ar analyses from corehole NRF 89-05 at 55 m BLS (Exp. nos.89-05_179a and 89-05_179b; table 1), and from corehole NRF 6P at 47 m BLS (Exp. nos. 6P_155a, b, and c; table 1). The sample from corehole NRF 89-05 yielded a weighted mean age of 712 ± 53 ka. The sample from corehole NRF 6P yielded a weighted mean age of 491 ± 30 ka.

The sample from NRF 6P at 47 m was also analyzed by using the $^{40}Ar/^{39}Ar$ method. The plateau age was 395 ± 25 ka (table 2; appendix B).

The 395 ± 25 ka age from the $^{40}Ar/^{39}Ar$ experiment seems reasonable (table 2). It is more than 200 k.y. younger than the oldest K-Ar age (Exp. no. 6P_155a; table 1) on the same sample. The K-Ar weighted mean age from the sample from corehole NRF 89-05 from 55 m BLS derived from two extractions (Exp. nos.89-05_179a and 89-05_179b, table 1) seems too old at 712 ± 53 ka. The NRF area is underlain by four more normal polarity flows and, thus, is much younger in age than the 0.78-Ma boundary of the Brunhes Normal Polarity Chron (fig. 3). This may be an example of an age that is too old owing to excess ^{40}Ar imbedded in phenocryst phases.

Central Facilities Area (CFA) Buried Vent Flow Group

Two sets of K-Ar age experiments were conducted on two samples from the Central Facilities Area (CFA) Buried Vent flow group at the NRF (pl. 1). One K-Ar experiment was performed on a sample from corehole NRF 89-04 from 65 m BLS and yielded a potentially reasonable weighted mean age of 492 ± 56 ka (table 1), from two extractions, which yielded different results (Exp. nos. 89-04_212a and 89-04_212b; table 1). The other K-Ar experiment was conducted on a

sample from corehole NRF 89-05 from 72 m BLS and yielded a weighted mean age of 598 ± 31 ka (table 1), from three extractions one of which (Exp. no. 89-05_235b) was much older than the other two extractions (Exp. nos. 89-05_235a and 89-05_235c; table 1).

The 492 ± 56 ka weighted mean age for the sample from corehole NRF 89-04 at 65 m is about 100 k.y. younger than the 617 ± 22 ka age for the CFA Buried Vent flow group in south INL coreholes that also were correlated by using paleomagnetic inclination (Champion and others, 2011). The 598 ± 31 ka weighted mean age is in much closer agreement. The more-than 280-k.y. span (Exp. no. 89-05_235b versus Exp. no. 89-04_212 b; table 1) of the five extraction ages for the samples from both coreholes (Exp. nos. 89-04_212a and b, 89-05_235 a, b, c), however, suggests caution in the use of this mean age.

State Butte Flow Group

The State Butte flow group (pl. 1) was sampled for a K-Ar experiment from corehole NRF 89-04 at 70 m BLS. The experiment yielded a weighted mean age of 819 ± 39 ka from two extractions (Exp. nos. 89-04_231a, b; table 1).

An $^{40}Ar/^{39}Ar$ age experiment was done on a sample from the State Butte flow group in the subsurface at the NRF. The experiment on a sample from corehole NRF 6P at 74 m BLS yielded a plateau age of 546 ± 47 ka (table 2; appendix B).

The K-Ar age (819 ± 39 ka; table 1) may indicate that the flow was emplaced during the Matuyama Reversed Polarity Chron (older than 0.78 Ma; fig. 3); the State Butte flow group, however, has normal magnetic polarity and is underlain by another normal polarity basalt flow group. Thus, the age is probably unreliable. The $^{40}Ar/^{39}Ar$ age of 546 ± 47 ka (table 2) seems to be reasonable.

Atomic Energy Commission (AEC) Butte Flow Group

The AEC Butte flow group (pl. 1) was sampled for a K-Ar age experiment from corehole NRF 6P from 113.4 m BLS. The K-Ar experiment yielded a weighted mean age of 384 ± 65 ka from three separate extractions which yielded different ages (Exp. nos. 6P_372 a, b, c; table 1).

An $^{40}Ar/^{39}Ar$ age experiment was also conducted on this sample. The $^{40}Ar/^{39}Ar$ age experiment yielded a plateau age of 727 ± 31 ka (table 2; appendix B).

The K-Ar weighted mean age of 384 ± 65 ka seems too young in stratigraphic context, and the $^{40}Ar/^{39}Ar$ plateau age of 727 ± 31 ka may be somewhat too old but is close to the 637 ± 35 ka age (Champion and others, 2011, p. 21) for this basalt flow group when uncertainties are taken into account.

North Late Matuyama Flow Group

The North Late Matuyama flow group (pl. 1) was sampled for two K-Ar age experiments from corehole NRF 7P at 134 m and 148 m BLS. The two K-Ar experiments, each from one extraction (Exp. nos. 7P_440 and 7P_486; table 1), yielded a weighted mean age of 604 ± 89 ka.

$^{40}Ar/^{39}Ar$ experiments were also performed on the same samples from corehole NRF 7P. These age experiments yielded plateau ages of 884 ± 53 ka and 1,176 ± 27 ka, respectively (table 2; appendix B).

The K-Ar weighted mean age of 604 ± 89 ka is too young because this reversed polarity basalt flow group must be older than 0.78 Ma (onset of the Brunhes Normal Polarity Chron) (fig. 3). The $^{40}Ar/^{39}Ar$ experiment that yielded the 1,176 ± 27 ka age seems too old because this flow is the shallowest Matuyama Reversed Polarity Chron flow in coreholes NRF 6P and NRF 7P and underlies flows of early Brunhes Normal Polarity Chron age (fig. 3; pl. 1). If the 1,176 ± 27 ka age were correct, then that would indicate an approximate 500-k.y. gap in basalt flow accumulation in the NRF area, which is unlikely (Champion and others, 2002). Additionally, this flow overlies a normal polarity flow that is 61 m deeper in corehole NRF 15 that was correlated to similar cored flows of Jaramillo Normal Polarity Subchron age farther to the south (Champion and others, 2011). The Jaramillo Normal Polarity Subchron (onset at 1.072 Ma, termination at 0.988 Ma) (fig. 3) is younger than 1,176 ± 27 ka, which suggests that this age experiment is not reasonable. The $^{40}Ar/^{39}Ar$ age of 884 ± 53 ka from corehole NRF 7P at 134 m BLS appears to be reasonable.

Stratigraphy

Plate 1 shows an interpretation of the subsurface stratigraphy in a south-to-north line of cross-section constructed through eight coreholes that span approximately 9 km (fig. 2). Paleomagnetic inclination and polarity determinations document that 13 basalt flow groups (pl. 1) were identified in coreholes USGS 133, NRF B18-1, NRF 89-05, NRF 89-04, NRF 6P, NRF 7P, NRF 15, and NRF 16. The basalt flow groups have normal and reversed magnetic polarities and were emplaced during the late Matuyama Reversed Polarity Chron, the Jaramillo Normal Polarity Subchron, and the Brunhes Normal Polarity Chron (fig. 3). Tie lines of correlation of sedimentary interbeds or the upper and lower surfaces of basalt flow groups were drawn between the coreholes on the basis of measured paleomagnetic inclination values (appendix A), polarity, and (or) age dates (tables 1 and 2) determined from samples of core.

Comparisons of paleomagnetic inclination and polarity measurements between samples from coreholes located only kilometers apart show comparable stratigraphic successions of mean inclination values with depth (pl. 1). At distances of greater than a few kilometers, correlation of mean inclination values is less consistent because basalt flow groups may be missing or additional basalt flow groups may be present and (or) found at different depth intervals. Some sedimentary interbeds found between some basalt flows and flow groups are correlative through parts of the cross-section; others pinch out between coreholes. The stratigraphy shown in the cross-section (pl. 1) is described from top to bottom and south to north. All coreholes began in surficial sediments of variable thickness. Surficial sediments in the NRF area are thinner than those found farther south at and near the INTEC, the CFA, and the RWMC (Anderson and others, 1996b).

West of Advanced Test Reactor Complex (ATRC) Flow Group

The West of ATRC flow group is found at the top of seven of the eight coreholes (USGS 133, NRF B18-1, NRF 89-05, NRF 6P, NRF 7P, NRF 15, and NRF 16) under sediments that range from about 3 to 10 m thick (pl. 1). Although it is also likely in corehole NRF 89-04, the upper 60 m of that hole was not sampled because of its close proximity to corehole NRF 89-05. The basalt flow group has mean inclination values that range from 73 to 77 degrees, is between about 15 m and 34 m thick, and is found as a single flow or multiple flows in the different coreholes. This basalt flow group is believed to have erupted from a vent to the west of the ATRC and is found in many southern INL coreholes (fig. 1; Champion and others, 2011). To date, no surface paleomagnetic sites in the flows of this vent have been sampled to compare with subsurface corehole inclination data.

Advanced Test Reactor Complex (ATRC) Unknown Vent Flow Group

The West of ATRC flow group is underlain by the ATRC Unknown Vent flow group (Champion and others, 2011). It is so labeled because it is thickest in coreholes near the ATRC (pl. 1) near corehole USGS 133. In corehole USGS 133, it has inclinations that range between 51 and 54 degrees separated by a petrographic boundary and unrecovered core (fig. 4; pl. 1; appendix A). In coreholes NRF B18-1, NRF 89-05, NRF 6P, NRF 7P, NRF 15, and NRF 16, it has mean inclinations that range from 55 to 58 degrees. A petrographic boundary similar to that in USGS 133 is found in coreholes NRF 89-05, NRF 6P, NRF 7P, and NRF 15 at similar depths when land surface elevation and the thickness of the overlying West of ATRC flow group are considered. This basalt flow group is also likely also in corehole NRF 89-04; the upper 60 m of that hole, however, was not sampled for paleomagnetic analysis. The ATRC Unknown Vent flow group is also found in coreholes near the INTEC area (fig. 12; Champion and others, 2011).

Champion and others (2011) correlated this basalt flow group with the North INTEC Buried Vent flow group in corehole NRF 7P. Paleomagnetic inclination values, stratigraphic position, and thickness in the north NRF area coreholes in this report, however, suggest that correlation of this basalt flow group to the ATRC Unknown Vent flow group is more probable (figs. 5–11; pl. 1).

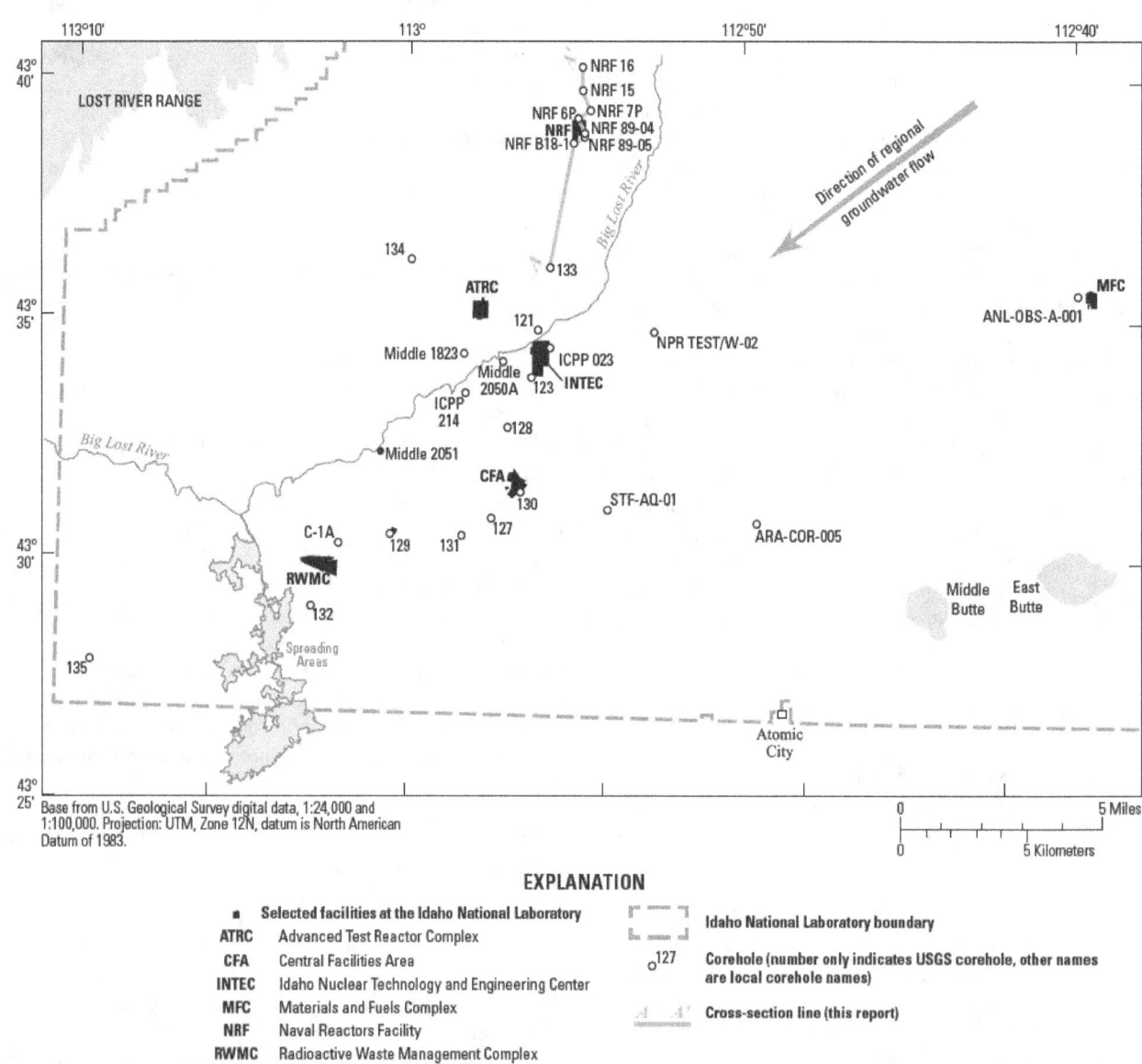

EXPLANATION

	Selected facilities at the Idaho National Laboratory
ATRC	Advanced Test Reactor Complex
CFA	Central Facilities Area
INTEC	Idaho Nuclear Technology and Engineering Center
MFC	Materials and Fuels Complex
NRF	Naval Reactors Facility
RWMC	Radioactive Waste Management Complex

Idaho National Laboratory boundary

127 Corehole (number only indicates USGS corehole, other names are local corehole names)

A   A' Cross-section line (this report)

Figure 12.   Location of coreholes in the southern part of the Idaho National Laboratory, Idaho (modified from Champion and others, 2011).

North Idaho Nuclear Technology and Engineering Center (INTEC) Buried Vent Flow Group

The North INTEC Buried Vent flow group underlies the ATRC Unknown Vent flow group and is found only in corehole USGS 133 in the NRF area (fig. 4; pl. 1). It has a mean inclination of 55 degrees and also is found in coreholes USGS 121, ICPP 023, USGS 123, and USGS 128 to the south near the INTEC (fig. 12; Champion and others, 2011). Champion and others (2011) correlated the North INTEC Buried Vent flow group to a 57-degree mean inclination basalt flow group in corehole NRF 7P. Data from the NRF area coreholes in this report suggest that correlation of the 57-degree basalt flow group to the ATRC Unknown Vent flow group is more probable.

Central Facilities Area (CFA) Buried Vent Flow Group

The CFA Buried Vent flow group underlies the North INTEC Buried Vent flow group in corehole USGS 133 (pl. 1) in the NRF area. It also is found in coreholes to the south near the INTEC (fig. 12; Champion and others, 2011). This basalt flow group is thickest at corehole USGS 128 about 2 km south of the INTEC and extends west to the RWMC and north to corehole NRF 7P (fig. 12; Champion and others, 2011). The CFA Buried Vent flow group has inclinations that range from 65 to 67 degrees in corehole USGS 133 (fig. 4; pl. 1). The CFA Buried Vent flow group in corehole USGS 133 likely correlates to basalt flow groups with mean inclinations that range from 62 to 67 degrees in coreholes to the north from NRF B18-1 to NRF 16 (pl. 1). This basalt flow group is thinner in the northern coreholes, which implies that the source is relatively far from the NRF area.

State Butte Flow Group

The State Butte flow group is found in five coreholes–NRF 89-04, NRF 6P, NRF 7P, NRF 15, and NRF 16 (pl. 1). The mean inclination values for this basalt flow group range from 71 to 74 degrees. These flows range from 44 to 59 m thick and thicken to the north, thus indicating a nearby source in that direction. Surface paleomagnetic sites at State Butte and adjacent surface basalt flows correlate well with this basalt flow group, and these flows probably erupted from the State Butte vent (fig. 1; Champion and others, 2011). The State Butte flow group probably pinches out to the south of corehole NRF 89-04 and is not found as far south as corehole USGS 133. State Butte flows may underlie coreholes NRF B18-1 and NRF 89-05 on the basis of their proximity to corehole NRF 89-04; these coreholes were not cored deeply enough to sample the State Butte flow group.

The State Butte flow group is the deepest cored basalt flow group in corehole NRF 89-04. In coreholes NRF 15 and NRF 16, the State Butte flow group is the oldest normal polarity basalt flow group emplaced after the Brunhes-Matuyama transition (fig. 3). In coreholes NRF 15 and NRF 16, reversed polarity basalt flow groups underlie the State Butte flow group.

Atomic Energy Commission (AEC) Butte Flow Group

The AEC Butte flow group is found in coreholes USGS 133, NRF 6P, and NRF 7P (pl. 1). This basalt flow group is likely in the subsurface near coreholes NRF B18-1, NRF 89-05, and NRF 89-04; these holes, however, were not cored deeply enough to sample it. The mean inclination values for this basalt flow group are in agreement at 57 degrees in coreholes NRF 6P and NRF 7P. In corehole USGS 133, the basalt flow group is thick and has mean inclinations that range from 50 to 54 degrees. Champion and others (2011) correlated this flow group to AEC Butte, which is found at the surface just north of the ATRC (fig. 1). The AEC Butte flow group also is found in the subsurface at the ATRC, the CFA, and the INTEC in coreholes USGS 134, USGS 133, USGS 121, ICPP 023, USGS 123, USGS 128, and Middle 1823 (Champion and others, 2011, pl. 1), and NPR Test/W-02 (fig. 12; Champion and others, 1988). Because the AEC Butte flow group is not found in coreholes NRF 15 and NRF 16, the northern edge of the basalt flow group is probably in the subsurface at or near corehole NRF 7P.

Late Basal Brunhes Flow Group

The Late Basal Brunhes flow group is found in corehole USGS 133, may be found in corehole NRF 7P, and does not correlate to any other coreholes in the NRF area (pl. 1). In corehole USGS 133, it has a mean inclination of 64 degrees. In corehole NRF 7P, the possible correlative basalt flow group has a mean inclination value of 63 degrees and is found below the AEC Butte flow group. The basalt flow group labeled the "Late Basal Brunhes flow group" in corehole NRF 7P (pl.1) could be a single flow from a different vent with similar mean inclination, or it could correlate to the Late Basal Brunhes flow group of Champion and others (2011) in coreholes ICPP 023, USGS 121, USGS 133, USGS 134, and Middle 1823 (fig. 12). That basalt flow group is found directly beneath flows from AEC Butte in coreholes proximal to the AEC Butte vent (fig. 1). The Late Basal Brunhes flow group is thickest in corehole USGS 134 west of the ATRC (fig. 12), and the vent is probably located in the subsurface near the ATRC (Champion and others, 2011). This flow is the deepest cored Brunhes Normal Polarity Chron (fig. 3) basalt flow group in coreholes NRF 7P and USGS 133.

East Matuyama Middle Flow Group

The East Matuyama Middle flow group underlies the State Butte flow group in coreholes NRF 15 and NRF 16 (pl. 1). It has reversed polarity and a mean inclination value of -42 degrees in corehole NRF 15, and -40 degrees in corehole NRF 16. This basalt flow group is the uppermost Matuyama Reversed Polarity Chron (fig. 3) basalt flow group in the NRF area and may correlate to the East Matuyama Middle flow group in coreholes NPR Test/W-02 and ANL-OBS-A-001 (fig. 12) to the south and east of the NRF area, which have similar inclinations (Champion and others, 2011).

North Late Matuyama Flow Group

The North Late Matuyama flow group is found in coreholes USGS 133, NRF 6P, NRF 7P, NRF 15, and NRF 16 (pl. 1). It has reversed magnetic polarity and mean inclination values that range from -46 to -48 degrees. Other NRF area coreholes were not cored deeply enough to sample this basalt flow group.

Additional Basalt Flow Groups and Sedimentary Interbeds Found at Depth in Corehole NRF 15

Corehole NRF 15 was cored more than 76 m deeper than other coreholes near the NRF and, therefore, penetrates additional basalt flow groups below the North Late Matuyama flow group. The stratigraphic order below the North Late Matuyama flow group is shown on plate 1. A 6-m-thick sedimentary interbed separates the North Late Matuyama flow group from the next underlying 4-m thick basalt, which has a mean inclination value of -58 degrees. This flow is underlain by a 5-m-thick sedimentary interbed that overlies a 2-m-thick basalt flow with a mean inclination of -69 degrees. The -69-degree basalt flow group possibly correlates to the south to the Matuyama flow group near the base of corehole USGS 133 that has -69-and -72-degree mean inclinations. The Matuyama flow group is extensive in the south INL and is thickest near the RWMC (Champion and others, 2011).

A 12-m- thick sedimentary interbed underlies the Matuyama flow group, which is in turn underlain by a 22-m-thick basalt flow group that comprises multiple basalt flows, has normal magnetic polarity, and has a mean inclination value of 54 degrees. This normal polarity basalt flow group probably correlates to the Jaramillo (Matuyama) flow group of Champion and others (2011, pl. 1), which has similar mean inclinations and is found in six coreholes of the south INL area

(USGS 134, Middle 2050A, Middle 1823, Middle 2051, C1A, and USGS 132) (fig. 12). This Jaramillo (Matuyama) flow group thickens to the south, which may indicate proximity to the eruptive vent (Champion and others, 2011, pl. 1). The Jaramillo (Matuyama) flow group in corehole NRF 15 now lies nearly 183 m (600 ft) higher in elevation in the NRF area than in the southern INL because of subsidence of the ESRP during the past 1 Ma (Champion and others, 2002).

The Jaramillo (Matuyama) normal polarity basalt flow group in corehole NRF 15 is underlain by a sedimentary interbed that overlies a reversed polarity basalt flow group that comprises two basalt flows with a mean inclination value of -58 degrees (pl. 1). The reversed polarity basalt flow group may correlate to the Matuyama 1.21-Ma flow group that has mean inclination values that range from -57 to -59 degrees and is found beneath the Jaramillo (Matuyama) flow group in coreholes Middle 1823, Middle 2050A, NPR Test/W-02, and ANL-OBS-A-001 (fig. 12) (Champion and others, 2011, pl. 1). This is the deepest cored basalt flow group in corehole NRF 15 and is underlain by more than 8.8 m of sedimentary interbeds.

Summary and Conclusions

Paleomagnetic inclination and polarity studies on samples from subsurface drill cores provide data that can help constrain the age and the extent of basalt flows and basalt flow groups. Data from paleomagnetic inclination and polarity studies were used to refine Pleistocene and Holocene stratigraphy at and near the Naval Reactors Facility (NRF) of the Idaho National Laboratory (INL).

Eight coreholes at and near the NRF were studied, and 749 subcores were collected and analyzed for paleomagnetic inclination and polarity. Samples of individual basalt flows from drill cores were collected in accordance with INL Lithologic Core Storage Library protocols. Paleomagnetic samples were collected from coreholes at depths from a few meters to 222 m.

Mean paleomagnetic inclination and polarity data from the samples were compared among the coreholes and basalt flow groups were determined. The basalt flow group determinations were then used to correlate subsurface basalt stratigraphy. Results demonstrate that stratigraphic successions of basalt flow groups correlate over distances of a few kilometers. Correlation of basalt flow groups in corehole USGS 133 to basalt flow groups to the north is less certain because it is more than 5 km from the nearest NRF area corehole.

Conventional K-Ar and ^{40}Ar/^{39}Ar age experiments were conducted on core samples and results were evaluated to determine the most reasonable ages and uncertainty levels for the different basalt flow groups of the NRF area. The selection of the most reasonable age for basalt flow groups that had multiple age experiments was based on stratigraphic position and paleomagnetic polarity.

Basalt shield volcanoes and their basalt flow groups have a wide range of eruptive volumes and areal extents. This study presented a detailed stratigraphic analysis of the NRF area and ties paleomagnetic data on these coreholes to earlier paleomagnetic data of other southern INL coreholes. A south-to-north cross-section shows correlation of basalt flow groups in the subsurface of the NRF area. Important correlations include the following:

- The West of Advanced Test Reactor Complex (ATRC) flow group is the uppermost basalt flow group in the NRF area and correlates among seven continuously cored holes in this study under surficial sediments. The West of ATRC flow group is also found in ATRC and INTEC coreholes and corehole USGS 129 (Champion and others, 2011).

- The ATRC Unknown Vent flow group correlates among seven continuously cored holes in this study that underly the West of ATRC flow group and a sedimentary interbed. Additional paleomagnetic inclination and stratigraphic data derived from the NRF coreholes changed the previously reported interpretation of the subsurface distribution of this basalt flow group (Champion and others, 2011). The ATRC Unknown Vent flow group also is found in coreholes near the ATRC and the INTEC.

- The Central Facilities Area (CFA) Buried Vent flow group correlates among all eight coreholes in the NRF area. It is also found in coreholes near the CFA and the Radioactive Waste Management Complex (RWMC) to the south. This basalt flow group is thickest near the CFA, which may indicate proximity to the vent. The State Butte flow group is found below the CFA Buried Vent flow group in the four northern NRF coreholes. It correlates to the State Butte surface vent located just northeast of the NRF. It is not found in coreholes south of the NRF.

- The Atomic Energy Commission (AEC) Butte flow group is found in coreholes USGS 133, NRF 6P, and NRF 7P. It probably underlies coreholes NRF B18-1,

NRF 89-05, and NRF 89-04, but those coreholes were not drilled deeply enough to penetrate the flow group. The AEC Butte flow group vent is exposed at the surface near the ATRC, and its flows are found in many coreholes near the ATRC and the INTEC (Champion and others, 2011). The AEC Butte flow group abruptly pinches out against the Matuyama Chron reversed polarity flows of the East Matuyama Middle flow group between coreholes NRF 7P and NRF 15.

- The East Matuyama Middle flow group correlates between coreholes NRF 15 and NRF 16 and may correlate to coreholes NPR Test/W-02 and ANL-OBS-A-001 (Champion and others, 2011).

- The North Late Matuyama flow group correlates among coreholes USGS 133, NRF 6P, NRF 7P, NRF 15, and NRF 16. It probably underlies coreholes NRF B18-1, NRF 89-05, and NRF 89-04, but those coreholes were not drilled deeply enough to penetrate the flow group. The vent that produced the North Late Matuyama flow group may be located in the general NRF area because it is thickest near corehole NRF 6P.

- The Matuyama flow group is found in coreholes in the southern INL from south of the RWMC to corehole USGS 133 and may extend north to corehole NRF 15. The Matuyama flow group is thickest near the RWMC and thins to the north.

- The Jaramillo (Matuyama) flow group is found in corehole NRF 15, which is the deepest NRF corehole, and shows that the basalt flow group is thick in the subsurface at NRF. This flow group is thickest between the RWMC and the INTEC and thins towards the ATRC and the NRF.

Acknowledgments

Potassium measurements were made by L. Espos and S.T. Pribble; argon measurements and age calculations were done by J.C. Von Essen. Assistance in the paleomagnetic sampling was provided by Dean Miller, Kyle Champion, and Joel Robinson. We gratefully acknowledge their help. We also wish to thank our manuscript peer reviewers. Their comments helped to significantly clarify the report. Funding for this research was provided by the U.S. Department of Energy, through the USGS INL Project Office, and by the USGS Volcano Science Center.

References Cited

Ackerman, D.J., Rattray, G.W., Rousseau, J.P., Davis, L.C., and Orr, B.R., 2006, A conceptual model of ground-water flow in the eastern Snake River Plain aquifer at the Idaho National Laboratory and vicinity with implications for contaminant transport: U.S. Geological Survey Scientific Investigations Report 2006-5122 (DOE/ID-22198), 62 p.

Ackerman, D.J., Rousseau, J.P., Rattray, G.W., and Fisher, J.C., 2010, Steady-state and transient models of groundwater flow and advective transport, eastern Snake River Plain aquifer, Idaho National Laboratory and vicinity, Idaho: U.S. Geological Survey Scientific Investigations Report 2010-5123 (DOE/ID-22209), 220 p.

Anderson, S.R., 1991, Stratigraphy of the unsaturated zone and uppermost part of the Snake River Plain aquifer at the Idaho Chemical Processing Plant and Test Reactor Area, Idaho National Engineering Laboratory, Idaho: U.S. Geological Survey Water-Resources Investigations Report 91-4010 (DOE/ ID-22095), 71 p.

Anderson, S.R., Ackerman, D.J., Liszewski, M.J., and Freiburger, R.M., 1996a, Stratigraphic data for wells at and near the Idaho National Engineering Laboratory, Idaho: U.S. Geological Survey Open-File Report 96-248 (DOE/ID-22127), 27 p. and 1 diskette.

Anderson, S.R., and Bartholomay, R.C., 1995, Use of natural-gamma logs and cores for determining stratigraphic relations of basalt and sediment at the Radioactive Waste Management Complex, Idaho National Engineering Laboratory, Idaho: Journal of the Idaho Academy of Science, v. 31, no. 1, p. 1–10.

Anderson, S.R., and Bowers, Beverly, 1995, Stratigraphy of the unsaturated zone and uppermost part of the Snake River Plain aquifer at Test Area North, Idaho National Engineering Laboratory, Idaho: U.S. Geological Survey Water-Resources Investigations Report 95-4130 (DOE/ ID-22122), 47 p.

Anderson, S.R., Kuntz, M.A., and Davis, L.C., 1999, Geologic controls of hydraulic conductivity in the Snake River Plain aquifer at and near the Idaho National Engineering and Environmental Laboratory, Idaho: U.S. Geological Survey Water-Resources Investigations Report 99-4033 (DOE/ ID-22155), 38 p.

Anderson, S.R., and Lewis, B.D., 1989, Stratigraphy of the unsaturated zone at the Radioactive Waste Management Complex, Idaho National Engineering Laboratory, Idaho: U.S. Geological Survey Water-Resources Investigations Report 89-4065 (DOE/ID-22080), 54 p.

Anderson, S.R., and Liszewski, M.J., 1997, Stratigraphy of the unsaturated zone and the Snake River Plain aquifer at and near the Idaho National Engineering Laboratory, Idaho: U.S. Geological Survey Water-Resources Investigations Report 97-4183 (DOE/ID-22142), 65 p.

Anderson, S.R., Liszewski, M.J., and Ackerman, D.J., 1996b, Thickness of surficial sediment at and near the Idaho National Engineering Laboratory, Idaho: U.S. Geological Survey Open-File Report 96-330 (DOE/ID-22128), 16 p.

Champion, D.E., Dalrymple, G.B., and Kuntz, M.A., 1981, Radiometric and paleomagnetic evidence for the Emperor reversed polarity event at 0.46 ± 0.05 m.y. in basalt lava flows from the eastern Snake River Plain, Idaho: Geophysical Research Letters, v. 8, no. 10, p. 1,055–1,058.

Champion, D.E., and Herman, T.C., 2003, Paleomagnetism of basaltic lava flows in coreholes ICPP-213, ICPP-214, ICPP-215, and USGS 128 near the Vadose Zone Research Park, Idaho Nuclear Technology and Engineering Center, Idaho National Engineering and Environmental Laboratory, Idaho: U.S. Geological Survey Open-File Report 2003-483 (DOE/ ID 22189), 21 p.

Champion, D.E., Hodges, M.K.V., Davis L.C., Lanphere, M.A., 2011, Paleomagnetic correlation of surface and subsurface basaltic lava flows and flow groups in the southern part of the Idaho National Laboratory, Idaho, with paleomagnetic data tables for drill cores: U.S. Geological Survey Scientific Investigations Report 2011-5049 (DOE/ ID-22214), 34 p.

Champion, D.E., Lanphere, M.A., Anderson. S.R., and Kuntz, M.A., 2002, Accumulation and subsidence of late Pleistocene basaltic lava flows of the eastern Snake River Plain, Idaho: Geological Society of America Special Paper 353, p. 175–192.

Champion, D.E., Lanphere, M.A., and Kuntz, M.A., 1988, Evidence for a new geomagnetic reversal from lava flows in Idaho: Discussion of short polarity reversals in the Brunhes and late Matuyama Polarity Chrons: Journal of Geophysical Research, v. 93, p. 11,667–11,680.

Champion, D.E., and Shoemaker, E. M., 1977, Paleomagnetic evidence for episodic volcanism on the Snake River Plain, Planetary Geology Field Conference on the Snake River Plain, Idaho, October 1977, NASA Technical Memorandum 78436, p.7–8.

Cox, A.V., and Dalrymple, G.B., 1967, Statistical analysis of geomagnetic reversal data and the precision of potassium-argon dating: Journal of Geophysical Research, v. 72, p. 2,603–2,614.

Dalrymple, G.B., and Duffield, W., 1988, High-precision $^{40}Ar/^{39}Ar$ dating of Oligocene rhyolites from the Mogollon-Datil Volcanic Field using a continuous laser system: Geophysical Research Letters, v. 15, p. 463–466.

Dalrymple, G.B., and Lanphere, M.A., 1969, Potassium-argon dating: W.H. Freeman, New York, 258 p.

Davis, L.C., Hannula, S.R., and Bowers, Beverly, 1997, Procedures for use of, and drill cores and cuttings available for study at the Lithologic Core Storage Library, Idaho National Engineering Laboratory, Idaho: U.S. Geological Survey Open-File Report 97-124 (DOE/ID-22135), 31 p.

Grimm-Chadwick, Claire, 2004, Petrogenesis of an evolved olivine tholeiite and chemical stratigraphy of cores USGS 127, 128, and 129, Idaho National Engineering and Environmental Laboratory: Idaho State University, Master's Thesis, 100 p. and appendixes.

Hackett, W.R., and Smith, R.P., 1992, Quaternary volcanism, tectonics, and sedimentation in the Idaho National Engineering Laboratory area, *in* Wilson, J.R., ed., Field guide to geologic excursions in Utah and adjacent areas of Nevada, Idaho, and Wyoming: Utah Geological Survey Miscellaneous Publication 92-3, p. 1–18.

Ingamells, C.O., 1970, Lithium metaborate flux in silicate analysis: Analytica Chimica Acta, v. 52, p. 323–334.

Kuntz, M.A., 1978, Geology of the Big Southern Butte area, eastern Snake River Plain, and potential volcanic hazards to the Radioactive Waste Management Complex, and other waste storage and reactor facilities at the Idaho National Engineering Laboratory, Idaho, with a section on Statistical treatment of the age of lava flows, by John O. Kork: U.S. Geological Survey Open-File Report 78-691, 70 p.

Kuntz, M.A., Anderson, S.R., Champion, D.E., Lanphere, M.A., and Grunwald, D.J., 2002, Tension cracks, eruptive fissures, dikes and faults related to late Pleistocene-Holocene basaltic volcanism and implications for the distribution of hydraulic conductivity in the eastern Snake River Plain, Idaho: Geological Society of America Special Paper 353, p. 111–133.

Kuntz, M.A., Covington, H.R., and Schorr, L.J., 1992, An overview of basaltic volcanism of the eastern Snake River Plain, Idaho, in Link, P.K., Kuntz, M.A., and Platt, L.B., Regional geology of eastern Idaho and western Wyoming: Geological Society of America Memoir 179, p. 227–267.

Kuntz, M.A., Dalrymple, G.B., Champion, D.E., and Doherty, D.J., 1980, An evaluation of potential volcanic hazards at the Radioactive Waste Management Complex, Idaho National Engineering Laboratory, Idaho: U.S. Geological Survey Open-File Report 80-388, 63 p., 1 map.

Kuntz, M. A., Skipp, Betty, Champion, D. E., Gans, P. B., Van Sistine, Paco, and Snyder, S. R., 2007, Geologic map of the Craters of the Moon 30' X 60' quadrangle, Idaho: U.S. Geological Survey Scientific Investigations Map 2969, 64-p. pamphlet, 1 plate, scale 1:100,000.

Kuntz, M.A., Skipp, Betty, Lanphere, M.A., Scott, W.E., Pierce, K.L., Dalrymple, G.B., Champion, D.E., Embree, G.F., Page W.R., Morgan, L.A., Smith, R.P., Hackett, W.R., and Rodgers, D.W., 1994, Geologic map of the Idaho National Engineering Laboratory and adjoining areas, eastern Idaho: U.S. Geological Survey Miscellaneous Investigations Map I-2330, scale 1:100,000.

Kuntz, M.A., Spiker, E.C., Rubin, Meyer, Champion, D.E., and Lefebvre, R.H., 1986, Radiocarbon studies of latest Pleistocene and Holocene lava flows of the Snake River Plain, Idaho—Data, lessons, interpretations: Quaternary Research, v. 25, no. 2, p. 163–176.

Lanphere, M.A., 2000, Comparison of conventional K-Ar and $^{40}Ar/^{39}Ar$ dating of young mafic volcanic rocks: Quaternary Research, v. 53, p. 294–301.

Lanphere, M.A., Champion, D.E., and Kuntz, M.A., 1993, Petrography, age, and paleomagnetism of basalt lava flows in coreholes Well 80, NRF 89-04, NRF 89-05, and ICPP 123, Idaho National Engineering Laboratory: U.S. Geological Survey Open-File Report 93-0327, 40 p.

Lanphere, M.A., Kuntz, M.A., and Champion, D.E., 1994, Petrography, age, and paleomagnetism of basaltic lava flows in coreholes at Test Area North (TAN), Idaho National Engineering Laboratory: U.S. Geological Survey Open-File Report 94-0686, 49 p.

Mankinen, E.A., and Dalrymple, G.B., 1972, Electron microprobe evaluation of terrestrial basalts for whole-rock dating: Earth and Planetary Science Letters, v. 17, issue 1, p. 89–94.

Mazurek, John, 2004, Genetic controls on basalt alteration within the eastern Snake River Plain aquifer system, Idaho: Idaho State University, Master's Thesis, 132 p. and appendixes.

McFadden, P.L., and Reid, A.B., 1982, Analysis of paleomagnetic inclination data: Journal of the Royal Astronomical Society, v. 69, issue 2, p. 307–319.

Miller, M.L., 2007, Basalt stratigraphy of corehole USGS-132 with correlations and petrogenetic interpretations of the B Flow Group, Idaho National Laboratory, Idaho: Idaho State University, Master's Thesis, 69 p., appendixes, and 1 pl.

Morse, L.H., and McCurry, M.O., 2002, Genesis of alteration of Quaternary basalts within a portion of the eastern Snake River Plain aquifer, *in* Link, P.K., and Mink, L.L., eds., Geology, hydrogeology, and environmental remediation—Idaho National Engineering and Environmental Laboratory, eastern Snake River Plain, Idaho: Boulder, Colo., Geological Society of America Special Paper 353, p. 213–224.

Ogg, J.G., and Smith, A.G., 2004, The geomagnetic polarity timescale, in Gradstein, F.M., Ogg, J.G., and Smith, A.G., eds., A geologic time scale 2004: Cambridge University Press, New York, 589 p.

Reed, M.F., Bartholomay, R.C., and Hughes, S.S., 1997, Geochemistry and stratigraphic correlation of basalt lavas beneath the Idaho Chemical Processing Plant, Idaho National Engineering Laboratory: Berlin, Germany, Environmental Geology, v. 30, no. 1-2, p. 108–118.

Rightmire, C.T., and Lewis, B.D, 1987, Geologic data collected and analytical procedures used during a geochemical investigation of the unsaturated zone, Radioactive Waste Management Complex, Idaho National Engineering Laboratory, Idaho: U.S. Geological Survey Open-File Report 87-0246 (DOE/ID-22072), 83 p.

Russell, I.C., 1902, Geology and water resources of the Snake River plains of Idaho: U.S. Geological Survey Series Bulletin Report no.199, 192 p.

Scarberry, K.C., 2003, Volcanology, geochemistry, and stratigraphy of the F Basalt Flow Group, eastern Snake River Plain, Idaho: Idaho State University, Master's Thesis, 139 p., 1 pl.

Shervais, J.W., Vetter, S.K., and Hanan, B.B., 2006, A layered mafic sill complex beneath the eastern Snake River Plain—Evidence from cyclic geochemical variations in basalt: Geology, v. 34, no. 5, p. 365–368.

Stacey, J.S., Sherrill, N.D., Dalrymple, G.B., Lanphere, M.A., and Carpenter, N.V., 1981, A five-collector system for the simultaneous measurement of argon isotope ratios in a static mass spectrometer: International Journal of Mass Spectrometry and Ion Physics, v. 39, p. 167–180.

Stout, M. Z., and Nicholls, J., 1977, Mineralogy and petrology of Quaternary lavas from the Snake River Plain, Idaho: Canadian Journal of Earth Sciences, v. 14, p. 2,140–2,156.

Tauxe, Lisa, Luskin, Casey, Selkin, Peter, Gans, Phillip, and Calvert, Andy, 2004, Paleomagnetic results from the Snake River Plain—Contribution to the time-averaged field global database: G3, Geochemistry, Geophysics, and Geosystems, American Geophysical Union and the Geochemical Society, v. 5, no. 8, 19 p.

Taylor, J.R., 1982, An introduction to error analysis: Mill Valley, CA, University Science Books, 270 p.

Twining, B.V., Hodges, M.K.V., and Orr, Stephanie, 2008, Construction diagrams, geophysical logs, and lithologic descriptions for boreholes USGS 126a, 126b, 127, 128, 129, 130, 131, 132, 133, and 134, Idaho National Laboratory, Idaho: U.S. Geological Survey Data Series Report 350 (DOE/ID-22205), 27 p., and appendixes.

Welhan, J.A., Johannesen, C.M., Davis, L.L., Reeves, K.S., and Glover, J.A., 2002, Overview and synthesis of lithologic controls on aquifer heterogeneity in the eastern Snake River Plain, Idaho, *in* Bonnichsen, Bill, White, C.M., and McCurry, M.O., eds., Tectonic and magmatic evolution of the Snake River Plain Volcanic Province: Idaho Geological Survey Bulletin 30, p. 435–460.

Wetmore, P.H., and Hughes, S.S., 1997, Change in magnitude of basaltic magmatism determined from model morphologies of subsurface quaternary basalts at the Idaho National Engineering and Environmental Laboratory, Idaho: Geological Society of America Abstracts with Programs, v. 29, no. 6, p. 365.

Wetmore, P.H., Hughes, S.S., Rodgers, D.W., and Anderson, S.R., 1999, Late Quaternary constructional development of the Axial Volcanic Zone, eastern Snake River Plain, Idaho: Geological Society of America Abstracts with Programs, v. 31, no. 4, p. 61.

Appendix A. Paleomagnetic Inclination Data for Selected Coreholes At and Near the Naval Reactors Facility, Idaho National Laboratory, Idaho

Table A1 contains depth and paleomagnetic inclination data from coreholes in cross-section A–A' presented in this report. The table presents inclination data that are organized alphabetically by corehole name.

Information in table A1 includes the corehole name, sample depth, and identification of sample groups that were used to determine the mean inclination for flows and 95-percent uncertainty. Each sample has a characteristic remanent inclination in degrees and an alternating field demagnetization level in milliTeslas or an alternative demagnetization approach. Positive inclination values indicate normal paleomagnetic polarity, and negative inclination values indicate reversed paleomagnetic polarity.

Petrographic boundaries denote a significant change in mineralogy in a sequence of flows. Unrecovered core and sediment intervals also were recorded in the depth column.

Some samples are labeled "NIIM", which stands for "Not Included In Mean." NIIM samples may have been thermally overprinted by overlying flows, tilted by endogenous inflation, struck by lightning when on the surface, or otherwise had their orientations disturbed so that they do not yield usable paleomagnetic inclination data.

Some samples carry the notation "therm," which is direction found by the difference vectors between sample magnetizations during the course of stepwise thermal demagnetizations. Stepwise thermal demagnetizations were done only on samples of particular interest.

The designation "TO" stands for "thermal overprinting." Samples that are thermally overprinted may require extra processing to discover their original orientation. If thermal overprinting was extensive, then the sample may not yield useful paleomagnetic data, and those samples were not included in means.

The "line fit" label means that the entire progressive alternating field demagnetization sequence was fit with a line on a vector component diagram to find the characteristic remanent inclination. Depth measurements are in feet and alternate field demagnetization level is in milliTeslas because that is how the raw data were collected.

Table A1.   Paleomagnetic inclination values for basalt samples from coreholes at and near the Naval Reactors Facility, Idaho National Laboratory, Idaho.

[Green shading marks start of the list of values for a different corehole. Grey shading delimits groups of samples used to determine the mean flow inclination and 95 percent uncertainty level. **Sample depth:** measured to nearest 1 foot. **Characteristic remanent inclination:** in degrees, down (negative value) or up (positive value) from the horizontal obtained by demagnetization. **Abbreviations**: ft, foot; AF, Alternating-Field; mT, milliTesla; Petrographic boundary, indicates noted change in mineralogy; NIIM, the sample was not included in the group mean inclination--see appendix explanation; Therm, stepwise thermal demagnetization was applied to the sample; TO, sample was thermally overprinted by overlying flow; line fit, the entire progressive alternating field demagnetization sequence was fit with a line on a vector component diagram to find the characteristic remanent inclination]

Sample depth (ft)	Characteristic remanent inclination (degrees)	AF demagnetization level (mT) or alternative demagnetization approach	Mean flow inclination and 95% uncertainty level for sample groupings (degrees)	Sample depth (ft)	Characteristic remanent inclination (degrees)	AF demagnetization level (mT) or alternative demagnetization approach	Mean flow inclination and 95% uncertainty level for sample groupings (degrees)
Corehole USGS 133				Corehole USGS 133—Continued			
28	72.3	20		213	57.9	30	
38	72.7	30		216	54.3	30	
47	74.5	30		223	51.8	30	
57	77.0	20		225	51.7	30	
67	76.6	20		228	54.1	30	
77	77.5	30		229–245	Unrecovered core and sediment		
86	75.6	30		247	54.3	30	
96	74.9	30		250	56.5	30	
106	73.6	20	74.6±1.1	252	53.5	20	
113	76.5	30		253	55.4	20	
120	71.6	30		255	56.4	30	
121–128	Sediments			258	59.9	20	
130	77.6	30		262	58.7	30	
132	74.7	30		266	56.6	30	
135	70.8	30		270	57.5	30	
137	75.0	40		272	55.6	30	
139	73.2	30		275	53.6	40	
				278	52.6	30	
141	52.5	30		282	54.2	30	55.2±1.0
145	55.3	20		287	53.5	30	
146	49.8	30		291	52.9	30	
148	50.3	40		294	53.3	30	
153	50.9	30		300	50.5	30	
154	51.8	20		302	53.4	30	
156	55.4	30		303	53.5	30	
158	54.2	30		306	55.2	30	
162	50.9	30		309	58.9	30	
165	52.8	30		312	56.7	30	
168	50.0	30	51.3±1.3	314	55.9	30	
170	47.1	30		315–317	Unrecovered core and petrographic boundary		
171	48.5	30		319	59.9	30	NIIM
174	53.3	30		324	63.9	30	
176	46.7	30		330	68.0	30	
178	45.4	20		334	69.7	30	67.3±2.2
179	51.5	30		338	67.1	30	
182	50.9	30		342	67.4	30	
185	53.0	30		346	67.6	30	
187	54.6	20		349	Petrographic boundary		
187–190	Unrecovered core and petrographic boundary			351	62.5	30	
192	52.8	30		354	64.0	30	
193	52.4	30		359	66.9	20	
195	56.1	30		364	65.4	30	65.1±1.7
201	51.1	30		369	64.1	30	
207	57.3	30		373	65.2	30	
211	59.2	30	54.4±1.9				

Table A1.   Paleomagnetic inclination values for basalt samples from coreholes at and near the Naval Reactors Facility, Idaho National Laboratory, Idaho.—Continued

[Green shading marks start of the list of values for a different corehole. Grey shading delimits groups of samples used to determine the mean flow inclination and 95 percent uncertainty level. **Sample depth:** measured to nearest 1 foot. **Characteristic remanent inclination:** in degrees, down (negative value) or up (positive value) from the horizontal obtained by demagnetization. **Abbreviations**: ft, foot; AF, Alternating-Field; mT, milliTesla; Petrographic boundary, indicates noted change in mineralogy; NIIM, the sample was not included in the group mean inclination--see appendix explanation; Therm, stepwise thermal demagnetization was applied to the sample; TO, sample was thermally overprinted by overlying flow; line fit, the entire progressive alternating field demagnetization sequence was fit with a line on a vector component diagram to find the characteristic remanent inclination]

Sample depth (ft)	Characteristic remanent inclination (degrees)	AF demagnetization level (mT) or alternative demagnetization approach	Mean flow inclination and 95% uncertainty level for sample groupings (degrees)	Sample depth (ft)	Characteristic remanent inclination (degrees)	AF demagnetization level (mT) or alternative demagnetization approach	Mean flow inclination and 95% uncertainty level for sample groupings (degrees)
Corehole USGS 133—Continued				Corehole USGS 133—Continued			
378	67.5	30		531	55.4	30	
379–382	Unrecovered core and sediment			537	55.5	30	
383	57.2	30	NIIM	544	54.9	30	
386	48.2	30		550	57.0	30	
388	51.1	20		556	53.4	30	
391	48.1	30	49.9±2.3	561	54.8	30	
393	48.1	20		564	50.6	20	
397	51.8	30		566	50.6	30	
398	52.3	30		567	Petrographic boundary		
400–402	Unrecovered core and petrographic boundary			569	53.9	30	
404	50.0	30					
409	49.5	30		575	59.5	20	NIIM
414	52.7	30		582	55.4	30	NIIM
420	53.9	30	51.6±1.6	588	61.9	30	
427	53.1	40		594	65.3	20	63.8±1.7
432	49.4	30		604	65.2	30	
437	53.1	30		610	64.1	30	
445	50.7	30		617	62.2	30	
447	Petrographic boundary			628	64.1	30	
448	55.2	20	NIIM	630–662	Unrecovered core and sediment		
452	45.5	20	NIIM	663	-42.0	50	NIIM
453	41.7	30	NIIM	665	-50.5	50	
455	39.9	20	NIIM	666	-48.5	30	
458	38.9	50	NIIM	669	-48.4	20	
462	45.5	20	NIIM	672	-48.4	30	
463	47.7	30	NIIM	676	-47.7	30	
467	50.6	50	NIIM	679	-45.8	30	-47.2±1.2
471	50.6	20	NIIM	683	-45.5	30	
475	55.3	30		687	-45.2	30	
479	54.1	30		689	-46.8	30	
482	56.9	30		692	-44.3	30	
486	53.5	30		695	-48.9	20	
488	46.9	20		697	-45.8	30	
492	59.4	30		698–700	Sediments		
494	53.6	30		701	-66.2	20	
497	56.3	30	54.4±1.1	704	-70.6	20	
501	55.3	20		708	-71.6	30	
503	55.1	10		712	-68.6	20	
505	53.8	20		717	-69.4	30	
510	53.6	20		720	-74.6	20	-71.6±2.6
513	57.0	30		723	-75.7	20	
520	52.3	30		724	Petrographic boundary		
525	56.2	30		726	-75.2	20	

Table A1.   Paleomagnetic inclination values for basalt samples from coreholes at and near the Naval Reactors Facility, Idaho National Laboratory, Idaho.—Continued

[Green shading marks start of the list of values for a different corehole. Grey shading delimits groups of samples used to determine the mean flow inclination and 95 percent uncertainty level. **Sample depth:** measured to nearest 1 foot. **Characteristic remanent inclination:** in degrees, down (negative value) or up (positive value) from the horizontal obtained by demagnetization. **Abbreviations:** ft, foot; AF, Alternating-Field; mT, milliTesla; Petrographic boundary, indicates noted change in mineralogy; NIIM, the sample was not included in the group mean inclination--see appendix explanation; Therm, stepwise thermal demagnetization was applied to the sample; TO, sample was thermally overprinted by overlying flow; line fit, the entire progressive alternating field demagnetization sequence was fit with a line on a vector component diagram to find the characteristic remanent inclination]

Sample depth (ft)	Characteristic remanent inclination (degrees)	AF demagnetization level (mT) or alternative demagnetization approach	Mean flow inclination and 95% uncertainty level for sample groupings (degrees)	Sample depth (ft)	Characteristic remanent inclination (degrees)	AF demagnetization level (mT) or alternative demagnetization approach	Mean flow inclination and 95% uncertainty level for sample groupings (degrees)
	Corehole USGS 133—Continued				Corehole NRF B18-1—Continued		
730	-72.4	20		154	51.9	50	
				159	56.7	50	
733	-67.5	20		163	55.6	40	
737	-70.6	20		167	54.6	40	
740	-66.8	30		170	56.4	50	
743	-71.3	20		175	55.7	40	
746	-67.7	30		177	45.4	30	
750	-68.9	20	-68.9±0.9	179	58.9	30	
754	-67.7	20		182	57.1	30	
759	-67.9	20		187	56.9	20	
764	-69.7	20		187–190	Probably sediment		
771	-68.8	20		195	67.4	20	
776	-68.9	20		198	66.5	20	
782	-70.5	20		203	65.3	20	
	Corehole NRF B18-1			209	63.5	20	64.8±1.8
				213	62.4	30	
39	75.2	20		219	63.0	30	
47	75.1	20		221	65.7	30	
55	76.2	20					
61	77.8	20	76.9±1.3	225	68.6	30	
67	76.2	20		227	68.2	20	
74	78.9	20		231	71.0	30	
81	77.6	40		233	67.1	30	
88	Petrographic boundary			236	61.1	40	66.9±2.4
93	78.4	20		240	68.7	40	
				242	67.7	30	
97	59.0	40		244	64.1	20	
100	54.5	40		246	62.5	30	
104	53.2	30		249	70.0	40	
107	59.0	30			Corehole NRF 89-05		
111	54.3	40					
114	55.6	20		22	45.6	20	NIIM
117	50.1	30		27	45.5	20	NIIM
119	50.5	30		32	68.6	30	
121	55.3	30		35	62.5	10	NIIM
123	54.2	30		39	74.9	10	
127	52.4	30		44	74.3	20	
130	52.3	30		54	73.9	20	73.3±1.5
136	55.7	30	54.8±1.2	64	72.9	20	
137	56.2	40		75	74.8	20	
141	55.5	40		85	71.2	10	
145	56.2	30		91	75.0	10	
150	55.4	30		99	74.3	20	

Table A1.   Paleomagnetic inclination values for basalt samples from coreholes at and near the Naval Reactors Facility, Idaho National Laboratory, Idaho.—Continued

[Green shading marks start of the list of values for a different corehole. Grey shading delimits groups of samples used to determine the mean flow inclination and 95 percent uncertainty level. **Sample depth:** measured to nearest 1 foot. **Characteristic remanent inclination:** in degrees, down (negative value) or up (positive value) from the horizontal obtained by demagnetization. **Abbreviations:** ft, foot; AF, Alternating-Field; mT, milliTesla; Petrographic boundary, indicates noted change in mineralogy; NIIM, the sample was not included in the group mean inclination--see appendix explanation; Therm, stepwise thermal demagnetization was applied to the sample; TO, sample was thermally overprinted by overlying flow; line fit, the entire progressive alternating field demagnetization sequence was fit with a line on a vector component diagram to find the characteristic remanent inclination]

Sample depth (ft)	Characteristic remanent inclination (degrees)	AF demagnetization level (mT) or alternative demagnetization approach	Mean flow inclination and 95% uncertainty level for sample groupings (degrees)	Sample depth (ft)	Characteristic remanent inclination (degrees)	AF demagnetization level (mT) or alternative demagnetization approach	Mean flow inclination and 95% uncertainty level for sample groupings (degrees)
Corehole NRF 89-05—Continued				Corehole NRF 89-04—Continued			
106	73.5	20		231	69.4	20	
107–119	Sediment			234	72.1	20	71.2±2.1
122	56.8	20		237	74.9	20	
125	54.3	20		241	70.9	20	
129	56.2	20		244	70.3	20	
131	55.3	20		247	74.3	20	
137	56.8	30		Corehole NRF 6P			
144	56.0	30					
150	54.7	30		14	71.0	20	
156	54.8	20		25	72.0	20	
157	Petrographic boundary		55.6±0.7	34	73.9	20	
159	44.3	30	NIIM	45	78.3	30	
162	47.3	20	NIIM	54	76.7	30	
165	52.3	20		64	79.1	20	75.3±2.0
170	56.6	20		73	78.2	30	
173	57.8	20		84	74.3	30	
179	55.7	20		93	75.2	30	
182	55.4	20		94–95	Sediment		
188	56.2	20		98	74.0	30	
191	55.7	20					
192–196	Sediment			103	67.6	30	TO
198	60.9	20		108	66.8	30	TO
202	59.5	20	61.5±17.0	114	60.1	40	TO
208	64.0	10					
				120	54.5	20	
212	65.3	20		125	55.8	30	
216	64.2	10		132	57.0	20	
221	63.9	20		138	54.1	10	
225	71.0	20	66.5±3.0	146	53.8	30	54.6±1.1
230	66.5	20		155	54.7	20	
235	64.1	20		161	52.8	30	
238	70.6	20		165	Petrographic boundary		
Corehole NRF 89-04				169	54.2	20	
192–196	Sediment			173	63.8	30	
198	63.3	20		179	64.5	30	
201	66.0	10		186	63.9	30	
204	62.8	20	66.1±3.0	191	64.4	30	64.8±1.5
208	67.0	20		199	64.6	30	
212	68.1	20		206	67.4	30	
217	69.3	10		210	Petrographic boundary		
218–221	0	Sediment		215	61.0	20	NIIM
223	67.6	10		219	70.1	20	
227	70.1	10		223	66.4	30	

Table A1. Paleomagnetic inclination values for basalt samples from coreholes at and near the Naval Reactors Facility, Idaho National Laboratory, Idaho.—Continued

[Green shading marks start of the list of values for a different corehole. Grey shading delimits groups of samples used to determine the mean flow inclination and 95 percent uncertainty level. **Sample depth:** measured to nearest 1 foot. **Characteristic remanent inclination:** in degrees, down (negative value) or up (positive value) from the horizontal obtained by demagnetization. **Abbreviations**: ft, foot; AF, Alternating-Field; mT, milliTesla; Petrographic boundary, indicates noted change in mineralogy; NIIM, the sample was not included in the group mean inclination--see appendix explanation; Therm, stepwise thermal demagnetization was applied to the sample; TO, sample was thermally overprinted by overlying flow; line fit, the entire progressive alternating field demagnetization sequence was fit with a line on a vector component diagram to find the characteristic remanent inclination]

Sample depth (ft)	Characteristic remanent inclination (degrees)	AF demagnetization level (mT) or alternative demagnetization approach	Mean flow inclination and 95% uncertainty level for sample groupings (degrees)	Sample depth (ft)	Characteristic remanent inclination (degrees)	AF demagnetization level (mT) or alternative demagnetization approach	Mean flow inclination and 95% uncertainty level for sample groupings (degrees)
Corehole NRF 6P—Continued				Corehole NRF 6P—Continued			
229	67.5	20		437	-48.0	Therm	
235	70.2	30		449	-50.8	Therm	
241	66.2	30		454	-49.4	Therm	
247	73.0	30		459	-48.0	Therm	-46.9±1.7
253	73.6	30		467	-45.9	Therm	
259	70.2	30		474	-44.2	Therm	
265	72.1	20		481	-43.5	Therm	
272	72.0	20		492	-43.7	Therm	
280	72.4	30	71.3±1.0	499	-45.5	Therm	
284	71.7	30		Corehole NRF 7P			
289	73.0	30		27	74.8	20	
296	72.3	30		30	72.6	30	
305	68.9	30		32	76.5	20	
310	Petrographic boundary			39	77.1	30	
312	73.3	30		44	75.2	20	
317	70.8	30		49	74.4	20	
323	72.8	30		54	74.1	20	75.8±1.4
328	72.8	30		56	76.7	30	
333	70.5	30		62	79.6	30	
337	72.6	30		67	71.6	20	
342	72.1	30		76	77.3	20	
347	74.5	30		83	77.7	30	
352	73.1	20		89	77.4	20	
353–354	Sediment			92–108	Sediment		
356	66.3	30	NIIM	106	65.5	30	NIIM
360	62.9	50	NIIM	112	60.3	20	
364	61.0	30		115	54.7	20	
370	56.5	30		119	55.4	30	
372	57.0	30		127	56.8	30	
377	59.1	30		131	54.3	20	57.3±1.8
384	50.6	30		133	Petrographic boundary		
387	54.6	20	56.6±1.6	133	59.4	20	
391	54.6	30		137	59.6	30	
394	53.9	20		143	60.1	30	
397	56.8	30		147	57.7	20	
401	56.9	30		151-152	Sediment		
407	57.1	30		155	54.3	30	
409	55.9	30					
413	56.3	30		159	65.1	30	
416	61.6	20		165	68.6	30	67.1±13.4
418	Petrographic boundary			170	67.7	50	
427	-48.4	Therm					
433	-48.4	Therm		173	71.3	30	

Table A1. Paleomagnetic inclination values for basalt samples from coreholes at and near the Naval Reactors Facility, Idaho National Laboratory, Idaho.—Continued

[Green shading marks start of the list of values for a different corehole. Grey shading delimits groups of samples used to determine the mean flow inclination and 95 percent uncertainty level. **Sample depth:** measured to nearest 1 foot. **Characteristic remanent inclination:** in degrees, down (negative value) or up (positive value) from the horizontal obtained by demagnetization. **Abbreviations:** ft, foot; AF, Alternating-Field; mT, milliTesla; Petrographic boundary, indicates noted change in mineralogy; NIIM, the sample was not included in the group mean inclination--see appendix explanation; Therm, stepwise thermal demagnetization was applied to the sample; TO, sample was thermally overprinted by overlying flow; line fit, the entire progressive alternating field demagnetization sequence was fit with a line on a vector component diagram to find the characteristic remanent inclination]

Sample depth (ft)	Characteristic remanent inclination (degrees)	AF demagnetization level (mT) or alternative demagnetization approach	Mean flow inclination and 95% uncertainty level for sample groupings (degrees)	Sample depth (ft)	Characteristic remanent inclination (degrees)	AF demagnetization level (mT) or alternative demagnetization approach	Mean flow inclination and 95% uncertainty level for sample groupings (degrees)
	Corehole NRF 7P—Continued				Corehole NRF 7P—Continued		
175	73.2	30		386	59.7	20	
181	72.0	20		390	57.4	20	
190	69.7	30		394	58.5	30	
200	70.4	30		400	58.5	30	
210	71.6	20		403-404	Sediment		
220	71.7	20		405	63.9	20	
229	70.0	30		409	61.2	20	
240	69.7	30		412	61.8	30	63.0±2.9
251	66.5	30		416	67.2	20	
254	Petrographic boundary			420	60.1	30	
258	57.3	50		424	63.6		
261	65.4	30		425-428	Sediment		
265	76.6	20		430	-43.8	40	
270	73.6	30		435	-44.8	20	
275	71.9	30		440	-45.6	10	
281	74.1	20		447	-44.7	30	
287	74.5	20		456	-48.6	30	-46.2±1.4
291	74.1	20	72.6±1.0	463	-49.6	20	
294	76.8	20		470	-44.4	20	
300	74.1	30		476	-46.2	20	
301	76.3	30		486	-46.2	10	
306	76.4	30		496	-47.6	50	
308	Petrographic boundary				Corehole NRF 15		
309	72.8	30		17	75.5	20	
313	71.7	30		20	75.2	40	
318	69.8	30		23	75.6	40	
321	68.4	30		26	72.6	30	
326	71.7	20		27	80.6	40	
330	70.0	30		32	80.4	30	
336	77.8	30		35	78.1	30	
341	69.5	20		40	80.5	40	77.0±1.4
345	77.0	30		44	78.8	40	
350	76.8	30		48	77.9	30	
353	74.8	30		52	80.1	40	
356-358	Sediment			54	74.5	40	
360	72.4	20		58	67.7	40	NIIM
367	69.9	30		59	74.9	30	
				65	73.7	40	
370	60.2	20		68	76.8	40	
374	55.2	20		71	76.0	30	
379	53.9	20		71-76	Probably sediment		
382	50.9	30	56.8±2.8	78	56.0	40	

Table A1. Paleomagnetic inclination values for basalt samples from coreholes at and near the Naval Reactors Facility, Idaho National Laboratory, Idaho.—Continued

[Green shading marks start of the list of values for a different corehole. Gray shading delimits groups of samples used to determine the mean flow inclination and 95 percent uncertainty level. **Sample depth:** measured to nearest 1 foot. **Characteristic remanent inclination:** in degrees, down (negative value) or up (positive value) from the horizontal obtained by demagnetization. **Abbreviations**: ft, foot; AF, Alternating-Field; mT, milliTesla; Petrographic boundary, indicates noted change in mineralogy; NIIM, the sample was not included in the group mean inclination--see appendix explanation; Therm, stepwise thermal demagnetization was applied to the sample; TO, sample was thermally overprinted by overlying flow; line fit, the entire progressive alternating field demagnetization sequence was fit with a line on a vector component diagram to find the characteristic remanent inclination]

Sample depth (ft)	Characteristic remanent inclination (degrees)	AF demagnetization level (mT) or alternative demagnetization approach	Mean flow inclination and 95% uncertainty level for sample groupings (degrees)	Sample depth (ft)	Characteristic remanent inclination (degrees)	AF demagnetization level (mT) or alternative demagnetization approach	Mean flow inclination and 95% uncertainty level for sample groupings (degrees)
Corehole NRF 15—Continued				Corehole NRF 15—Continued			
81	56.6	30		214	77.3	30	73.4±1.2
86	57.8	40		216	75.4	40	
89	60.0	40		220	74.8	30	
92	61.6	40		225	73.5	40	
95	56.9	30		230	78.8	40	
97	58.4	30		235	69.4	40	
99	56.3	40		239	75.5	40	
103	56.3	30		243	74.8	40	
106	55.9	40		247	75.0	40	
108	57.9	40		251	70.7	30	
110	56.4	40	57.6±0.6	256	71.4	40	
113	59.6	30		260	71.5	30	
115	58.1	40		264	71.7	40	
118	58.0	30		266	Petrographic boundary		
126	56.3	40		269	55.8	40	NIIM
131	56.1	30		274	73.1	40	
133	56.4	40		276	72.6	40	
136	58.1	30		279	75.3	40	
138	58.0	30		283	76.6	40	
142	56.4	40		286	69.3	40	
143	59.3	40		289	71.9	40	
147	57.7	40		292	74.7	40	
150	58.0	40		296	74.1	40	
155	Petrographic boundary			300	74.5	40	72.0±1.3
156	56.7	Therm		304	72.3	40	
				308	73.5	40	
160	64.4	Therm		310	69.3	40	
164	62.6	40		314	69.9	30	
169	65.3	30	64.1±1.2	317	72.5	40	
172	63.0	30		322	65.8	30	
174	64.8	40		326	68.9	40	
175	64.7	40		331	71.2	30	
176–181	Probably sediment			335	73.9	40	
182	73.8	30		337–341	Probably sediment		
184	68.0	40		344	-36.8	Therm	
186	72.2	40		348	-44.9	Therm	
190	70.7	40		353	-45.5	40	
192	61.9	40	NIIM	358	-47.8	40	
195	72.0	40		363	-46.5	30	
202	73.8	40		368	-36.5	40	
203	74.9	30		373	-44.7	40	
206	75.0	40		379	-45.7	40	
212	73.7	40		384	-44.7	30	-41.5±2.1
				389	-37.8	40	

Table A1.   Paleomagnetic inclination values for basalt samples from coreholes at and near the Naval Reactors Facility, Idaho National Laboratory, Idaho.—Continued

[Green shading marks start of the list of values for a different corehole. Gray shading delimits groups of samples used to determine the mean flow inclination and 95 percent uncertainty level. **Sample depth:** measured to nearest 1 foot. **Characteristic remanent inclination:** in degrees, down (negative value) or up (positive value) from the horizontal obtained by demagnetization. **Abbreviations**: ft, foot; AF, Alternating-Field; mT, milliTesla; Petrographic boundary, indicates noted change in mineralogy; NIIM, the sample was not included in the group mean inclination--see appendix explanation; Therm, stepwise thermal demagnetization was applied to the sample; TO, sample was thermally overprinted by overlying flow; line fit, the entire progressive alternating field demagnetization sequence was fit with a line on a vector component diagram to find the characteristic remanent inclination]

Sample depth (ft)	Characteristic remanent inclination (degrees)	AF demagnetization level (mT) or alternative demagnetization approach	Mean flow inclination and 95% uncertainty level for sample groupings (degrees)	Sample depth (ft)	Characteristic remanent inclination (degrees)	AF demagnetization level (mT) or alternative demagnetization approach	Mean flow inclination and 95% uncertainty level for sample groupings (degrees)
Corehole NRF 15—Continued				Corehole NRF 15—Continued			
392	-38.0	40		525	-57.9	line fit	
394	-38.3	40		526–542	Probably sediment		
397	-33.6	40		542	-65.2	line fit	
400	-35.7	40		545	-70.6	line fit	-68.5±5.0
403	-43.1	40		546	-68.9	line fit	
406	-39.6	40		548	-69.1	line fit	
409	-41.9	40		550–589	Probably sediment		
413	-44.4	40		590	47.3	line fit	NIIM
416	-42.8	40		594	51.1	line fit	
416–419	Probably sediment			598	53.0	line fit	
421	-50.9	40		603	53.7	line fit	
425	-47.3	30		608	48.1	line fit	NIIM
429	-46.0	40		609	52.7	line fit	
433	-44.4	40		610	49.3	line fit	NIIM
438	-47.9	30		614	50.7	line fit	
440	-45.5	40		619	56.6	line fit	
442	-46.2	40	-46.1±1.8	623	54.1	line fit	
445	-39.2	30		629	54.0	line fit	53.7±1.1
447	-43.4	40		633	54.9	line fit	
451	-47.5	40		637	53.2	line fit	
452	-50.5	30		641	48.8	line fit	NIIM
456	-45.6	40		642	48.5	line fit	NIIM
458	-45.4	40		643	47.4	line fit	NIIM
460	Petrographic boundary			646	49.0	line fit	NIIM
462	-44.5	40		648	53.8	line fit	
465	-46.9	40		650	58.0	line fit	
470	-49.9	30		654	53.8	line fit	
473	-53.4	30		656	51.6	line fit	
476	-46.4	30		657	Petrographic boundary		
480	-47.9	30		659	54.5	line fit	
483	-46.4	40	-47.5±1.5				
484	-44.6	40		663	-54.2	line fit	
485	-48.3	30		669	-58.2	line fit	
487	-46.1	40		675	-55.9	line fit	
488	-50.1	40		681	-56.7	line fit	
490	-46.5	30		684	-56.4	line fit	
491	-46.8	30		690	-58.7	line fit	-57.9±1.2
491–512	Probably sediment			696	-61.0	line fit	
514	-57.8	30		701	-59.5	line fit	
516	-56.8	line fit		709	-59.3	line fit	
519	-57.1	line fit		715	-58.3	line fit	
521	-57.8	line fit	-57.9±0.9	722	-58.6	line fit	
522	-59.6	line fit		727	-57.5	line fit	
524	-58.3	line fit		730–759	Probably sediment		

Table A1. Paleomagnetic inclination values for basalt samples from coreholes at and near the Naval Reactors Facility, Idaho National Laboratory, Idaho.—Continued

[Green shading marks start of the list of values for a different corehole. Gray shading delimits groups of samples used to determine the mean flow inclination and 95 percent uncertainty level. **Sample depth:** measured to nearest 1 foot. **Characteristic remanent inclination:** in degrees, down (negative value) or up (positive value) from the horizontal obtained by demagnetization. **Abbreviations**: ft, foot; AF, Alternating-Field; mT, milliTesla; Petrographic boundary, indicates noted change in mineralogy; NIIM, the sample was not included in the group mean inclination--see appendix explanation; Therm, stepwise thermal demagnetization was applied to the sample; TO, sample was thermally overprinted by overlying flow; line fit, the entire progressive alternating field demagnetization sequence was fit with a line on a vector component diagram to find the characteristic remanent inclination]

Sample depth (ft)	Characteristic remanent inclination (degrees)	AF demagnetization level (mT) or alternative demagnetization approach	Mean flow inclination and 95% uncertainty level for sample groupings (degrees)	Sample depth (ft)	Characteristic remanent inclination (degrees)	AF demagnetization level (mT) or alternative demagnetization approach	Mean flow inclination and 95% uncertainty level for sample groupings (degrees)
Corehole NRF 16				Corehole NRF 16—Continued			
16	32.9	line fit	NIIM	141	69.1	line fit	
18	68.6	line fit	NIIM	141–144	Petrographic boundary		
20	78.5	line fit		147	69.2	line fit	
26	75.2	line fit		152	70.0	line fit	
29	81.4	line fit		159	74.0	line fit	
32	73.9	line fit		167	72.3	line fit	
34	76.0	line fit		174	74.5	line fit	72.2±1.5
35	78.8	line fit		180	73.1	line fit	
37	73.9	line fit	75.8±1.4	187	70.6	line fit	
40	74.4	line fit		192	71.1	line fit	
43	74.5	line fit		199	75.6	line fit	
46	81.0	line fit		206	71.7	line fit	
49	77.7	line fit		208–210	Petrographic boundary		
51	73.9	line fit		212	73.3	line fit	
53	75.7	line fit		216	73.5	line fit	
55	77.9	line fit		220	70.4	line fit	
56	77.0	line fit		224	73.0	line fit	
64	72.7	line fit		227	73.6	line fit	
65–69	Probably sediment			228	75.1	line fit	
69	-15.2	line fit	NIIM	230	72.2	line fit	
72	20.7	line fit	NIIM	232	72.3	line fit	
75	45.1	line fit	NIIM	235	75.3	line fit	
79	66.5	line fit	NIIM	237	71.7	line fit	
82	58.0	line fit		240	70.4	line fit	
89	51.9	line fit		245	70.7	line fit	
90	58.1	line fit		249	71.2	line fit	
92	58.1	line fit		254	72.3	line fit	
94	54.7	line fit		258	73.0	line fit	72.9±0.8
99	55.1	line fit	54.8±1.4	262	67.4	line fit	
102	55.7	line fit		264	75.5	line fit	
105	53.5	line fit		266	71.8	line fit	
110	56.1	line fit		269	76.6	line fit	
114	52.5	line fit		272	71.4	line fit	
117	53.9	line fit		275	72.4	line fit	
120	51.6	line fit		278	73.5	line fit	
65–69	Probably sediment			280	76.6	line fit	
125	52.5	Therm		284	75.3	line fit	
				289	72.0	line fit	
128	61.0	line fit		291	70.2	line fit	
130	62.3	line fit		293	74.1	line fit	
131	59.2	line fit		294	74.4	line fit	
134	62.0	line fit	64.0±3.9	296	73.8	line fit	
136	69.6	line fit		299	75.2	line fit	
138	64.9	line fit		300	Petrographic boundary		

Table A1.   Paleomagnetic inclination values for basalt samples from coreholes at and near the Naval Reactors Facility, Idaho National Laboratory, Idaho.—Continued

[Green shading marks start of the list of values for a different corehole. Gray shading delimits groups of samples used to determine the mean flow inclination and 95 percent uncertainty level. **Sample depth:** measured to nearest 1 foot. **Characteristic remanent inclination:** in degrees, down (negative value) or up (positive value) from the horizontal obtained by demagnetization. **Abbreviations**: ft, foot; AF, Alternating-Field; mT, milliTesla; Petrographic boundary, indicates noted change in mineralogy; NIIM, the sample was not included in the group mean inclination--see appendix explanation; Therm, stepwise thermal demagnetization was applied to the sample; TO, sample was thermally overprinted by overlying flow; line fit, the entire progressive alternating field demagnetization sequence was fit with a line on a vector component diagram to find the characteristic remanent inclination]

Sample depth (ft)	Characteristic remanent inclination (degrees)	AF demagnetization level (mT) or alternative demagnetization approach	Mean flow inclination and 95% uncertainty level for sample groupings (degrees)	Sample depth (ft)	Characteristic remanent inclination (degrees)	AF demagnetization level (mT) or alternative demagnetization approach	Mean flow inclination and 95% uncertainty level for sample groupings (degrees)
Corehole NRF 16—Continued				Corehole NRF 16—Continued			
302	70.1	line fit		366	-37.7	line fit	
305	74.0	line fit		369	-40.0	line fit	
309	72.8	line fit		371	-45.8	line fit	
312	75.8	line fit		374	-43.8	line fit	
319	74.9	line fit		376	-41.4	line fit	
321	77.4	line fit	73.8±1.5	378	-38.5	line fit	
322	76.3	line fit		378–381	Probably sediment		
324	76.2	line fit		382	-38.8	Therm	
326	75.5	line fit					
329	72.7	line fit		386	-45.9	line fit	
332	70.7	line fit		390	-47.3	line fit	
336	70.8	line fit		392	-67.0	line fit	NIIM
337	Sediment			397	-50.4	line fit	
339	-35.3	Therm		399	-46.1	line fit	
342	-37.7	Therm		401	-48.3	line fit	
345	-45.8	Therm		404	-49.3	line fit	-47.3±1.4
348	-41.2	line fit		407	-50.8	line fit	
350	-38.4	line fit		409	-47.2	line fit	
353	-34.4	Therm		413	-44.1	line fit	
356	-44.3	Therm		417	-44.5	line fit	
359	-38.8	line fit		420	-47.6	line fit	
362	-42.8	line fit	-40.3±1.7	424	-45.5	line fit	
364	-40.1	line fit					

Appendix B. ⁴⁰Ar/³⁹Ar Analytical Data for Selected Coreholes, Naval Reactors Facility, Idaho National Laboratory, Idaho

Table B1 shows analytical data associated with determination of the plateau ages of samples from coreholes NRF 6P and 7P, Naval Reactors Facility, Idaho National Laboratory, Idaho.

J is the neutron fluence constant calculated from the irradiation of a known age monitor mineral. Plateau age is the weighted mean age using [n of m] diffusion steps. The isochron age is derived from the slope of the best fit line on an isochron diagram plotting $^{40}Ar/^{36}Ar$ versus $^{39}Ar/^{36}Ar$ isotopic ratios; Inverse isochron age is derived from the slope of the best fit line on an isochron diagram plotting $^{36}Ar/^{40}Ar$ versus $^{39}Ar/^{40}Ar$ isotopic ratios; total gas age is calculated by summing all diffusion steps.

Temp (°C) is the temperature of each progressive argon diffusion step. $^{40}Ar/^{39}Ar$, $^{37}Ar/^{39}Ar$, and $^{36}Ar/^{39}Ar$ are the ratios of the abundances of argon isotopes ^{40}Ar, ^{39}Ar, ^{37}Ar, and ^{36}Ar. Moles $^{40}Ar_{rad}$ is the absolute number of moles of ^{40}Ar measured in each diffusion step. $^{40}Ar_{rad}$ (%) is the percentage of radiogenic ^{40}Ar measured in each step; $^{39}Ar_{Ca}$ (%) and $^{36}Ar_{Ca}$ (%) are the percentages of ^{39}Ar and ^{36}Ar correcting those isotope abundances for calcium interference for each step. K/Ca is the potassium/calcium ratio calculated for each step from the ^{39}Ar and ^{37}Ar isotope measurements. ^{39}Ar (%) is the percentage of ^{39}Ar released (100% total) during each step. Age±s.d. (Ma) is the age and standard deviation in age, in millions of years, calculated for each step.

Figures in appendix B show the forward and inverse isochron diagrams and argon age plateau diagrams for each basalt sample analyzed by the $^{40}Ar/^{39}Ar$ method.

Isochron Diagrams

Forward isochron diagrams plot the $^{40}Ar/^{36}Ar$ ratio against the $^{39}Ar/^{36}Ar$ ratio for each temperature step of the argon age experiment. The inverse isochron diagram plots the $^{36}Ar/^{40}Ar$ ratio against the $^{39}Ar/^{40}Ar$ ratio for each temperature step of the argon age experiment. The age of the isochron with uncertainty, $^{40}Ar/^{36}Ar$ intercept value with uncertainty, and Mean Square of Weighted Deviates (Sums/(N-2)) are also shown. Open circle data points were used in the isochron fit, cross data points were not used.

Argon Age Plateau Diagrams

Argon age plateau diagrams plot the age of the sample in millions of years (Ma) against the cumulative percent of ^{39}Ar released in the argon age experiment. Rectangular boxes show the age and the 1 sigma uncertainty in age plotted against the span of ^{39}Ar released during each successive temperature step of the argon age experiment. The small italic numbers are temperatures (°C) of the progressive steps. The horizontal double-headed arrowed line with enclosed number shows the age and 1 sigma uncertainty in age, and the percent span of ^{39}Ar used in the weighted mean age calculation for this age experiment.

Table B1. $^{40}Ar/^{39}Ar$ analytical data, with plateau, forward and inverse isochron diagrams for basalt samples from coreholes NRF 6P and NRF 7P, Naval Reactors Facility, Idaho National Laboratory, Idaho.

[J is the neutron fluence constant calculated from the irradiation of a known age monitor mineral. Plateau age is the weighted mean age using [n of m] diffusion steps; Isochron age is derived from the slope of the best fit line on an isochron diagram plotting $^{40}Ar/^{36}Ar$ versus $^{39}Ar/^{36}Ar$ isotopic ratios; Inverse isochron age is derived from the slope of the best fit line on an isochron diagram plotting $^{36}Ar/^{40}Ar$ versus $^{39}Ar/^{40}Ar$ isotopic ratios; total gas age is calculated by summing all diffusion steps. **Temp (°C):** temperature of each progressive argon diffusion step; $^{40}Ar/^{39}Ar$, $^{37}Ar/^{39}Ar$, and $^{36}Ar/^{39}Ar$: ratios of the abundances of argon isotopes ^{40}Ar, ^{39}Ar, ^{37}Ar, and ^{36}Ar. **Moles $^{40}Ar_{rad}$:** absolute number of moles of ^{40}Ar measured in each diffusion step. $^{40}Ar_{rad}$ **(%):** percentage of radiogenic ^{40}Ar measured in each step. $^{39}Ar_{Ca}$ **(%)** and $^{36}Ar_{Ca}$ **(%):** percentages of ^{39}Ar and ^{36}Ar correcting those isotope abundances for calcium interference for each step. **K/Ca:** potassium/calcium ratio calculated for each step from the ^{39}Ar and ^{37}Ar isotope measurements. ^{39}Ar **(%):** percentage of ^{39}Ar released (100% total) during each step. **Age±s.d.:** age and standard deviation in age, in millions of years (Ma), calculated for each step]

NRF 6P (155 ft depth) [J=0.00038155]

Plateau age=0.395±0.025 Ma; isochron age=0.372±0.038 Ma; inverse isochron age=0.371±0.037 Ma; total gas age=0.382±0.028 Ma

Temp (C°)	$^{40}Ar/^{39}Ar$	$^{37}Ar/^{39}Ar$	$^{36}Ar/^{39}Ar$	Moles $^{40}Ar_{rad}$	$^{40}Ar_{rad}$ (%)	$^{39}Ar_{Ca}$ (%)	$^{36}Ar_{Ca}$ (%)	K/Ca	^{39}Ar (%)	Age±s.d. (Ma)
450	9.818	8.480	0.03308	1.701^{-14}	5.6	0.55	5.3	0.057	53.6	0.382 ± 0.041
500	6.176	6.000	0.02015	7.022^{-15}	9.4	0.39	6.1	0.081	21.0	0.402 ± 0.038
600	6.853	6.144	0.02240	2.728^{-15}	8.8	0.40	5.6	0.079	7.9	0.417 ± 0.078
650	12.318	5.371	0.04072	1.948^{-15}	4.9	0.35	2.7	0.091	5.6	0.420 ± 0.109
700	14.252	4.703	0.04792	9.497^{-16}	2.6	0.30	2.0	0.104	4.4	0.257 ± 0.136
750	33.28	5.951	0.11170	1.327^{-15}	1.9	0.38	1.1	0.082	3.7	0.435 ± 0.179
800	42.57	12.376	0.14233	1.314^{-15}	3.0	0.80	1.8	0.039	1.8	0.874 ± 0.340
900	66.98	64.36	0.2414	-5.334^{-16}	-0.7	4.2	5.5	0.007	2.0	-0.326 ± 0.413

NRF 6P (241 ft depth) [J=0.00037585]

Plateau age=0.546±0.047 Ma; isochron age=0.484±0.089 Ma; inverse isochron age=0.485±0.084 Ma; total gas age=0.561±0.065 Ma

Temp (C°)	$^{40}Ar/^{39}Ar$	$^{37}Ar/^{39}Ar$	$^{36}Ar/^{39}Ar$	Moles $^{40}Ar_{rad}$	$^{40}Ar_{rad}$ (%)	$^{39}Ar_{Ca}$ (%)	$^{36}Ar_{Ca}$ (%)	K/Ca	^{39}Ar (%)	Age±s.d. (Ma)
450	28.18	1.8562	0.09276	1.254^{-14}	3.1	0.014	0.12	0.264	50.2	0.596 ± 0.078
500	9.723	2.072	0.03107	2.249^{-15}	6.8	0.039	0.13	0.236	11.9	0.451 ± 0.100
575	12.900	8.698	0.04262	3.606^{-15}	6.4	0.030	0.56	0.056	15.2	0.565 ± 0.087
625	23.75	9.709	0.07940	1.911^{-15}	3.7	0.016	0.63	0.050	7.6	0.597 ± 0.164
675	35.92	12.040	0.12237	6.825^{-16}	1.4	0.011	0.78	0.040	4.9	0.336 ± 0.256
725	52.88	10.674	0.17381	1.742^{-15}	4.1	0.007	0.69	0.046	2.8	1.478 ± 0.431
900	198.49	123.02	0.6973	-1.971^{-16}	-0.1	0.002	7.9	0.0037	5.6	-0.840 ± 0.763

NRF 6P (372 ft depth) [J=0.00036836]

plateau age = 0.727 ± 0.031 Ma; isochron age = 0.752 ± 0.063 Ma; inverse isochron age = 0.754 ± 0.063 Ma; total gas age = 0.635 ± 0.077 Ma

Temp (C°)	$^{40}Ar/^{39}Ar$	$^{37}Ar/^{39}Ar$	$^{36}Ar/^{39}Ar$	Moles $^{40}Ar_{rad}$	$^{40}Ar_{rad}$ (%)	$^{39}Ar_{Ca}$ (%)	$^{36}Ar_{Ca}$ (%)	K/Ca	^{39}Ar (%)	Age±s.d. (Ma)
450	47.68	4.756	0.15898	3.923^{-14}	2.1	0.31	0.61	0.103	58.2	0.658 ± 0.130
500	16.082	0.95151	0.05072	1.503^{-14}	7.1	0.061	0.38	0.515	19.2	0.764 ± 0.042
550	12.936	7.491	0.04187	5.351^{-15}	7.8	0.48	3.7	0.065	7.7	0.677 ± 0.054
600	14.550	7.894	0.04717	2.846^{-15}	7.5	0.51	3.4	0.062	3.8	0.726 ± 0.087
700	20.58	11.058	0.07012	1.116^{-15}	2.5	0.71	3.2	0.044	3.1	0.351 ± 0.111
750	29.46	10.879	0.09960	8.001^{-16}	2.3	0.70	2.2	0.045	1.7	0.460 ± 0.187
800	37.65	11.369	0.12872	2.998^{-16}	0.8	0.73	1.8	0.043	1.5	0.202 ± 0.222
850	72.22	17.075	0.2444	1.220^{-15}	1.4	1.1	1.4	0.028	1.7	0.691 ± 0.247
900	120.57	84.03	0.4274	-1.125^{-15}	-0.5	5.4	4.0	0.0055	2.4	-0.451 ± 0.462
950	114.13	102.02	0.4048	3.099^{-16}	0.6	6.6	5.2	0.0045	0.6	0.485 ± 0.668

Table B1. $^{40}Ar/^{39}Ar$ analytical data, with plateau, forward and inverse isochron diagrams for basalt samples from coreholes NRF 6P and NRF 7P, Naval Reactors Facility, Idaho National Laboratory, Idaho.—Continued

[J is the neutron fluence constant calculated from the irradiation of a known age monitor mineral. Plateau age is the weighted mean age using [n of m] diffusion steps; Isochron age is derived from the slope of the best fit line on an isochron diagram plotting $^{40}Ar/^{36}Ar$ versus $^{39}Ar/^{36}Ar$ isotopic ratios; Inverse isochron age is derived from the slope of the best fit line on an isochron diagram plotting $^{36}Ar/^{40}Ar$ versus $^{39}Ar/^{40}Ar$ isotopic ratios; total gas age is calculated by summing all diffusion steps. **Temp (°C):** temperature of each progressive argon diffusion step; $^{40}Ar/^{39}Ar$, $^{37}Ar/^{39}Ar$, and $^{36}Ar/^{39}Ar$: ratios of the abundances of argon isotopes ^{40}Ar, ^{39}Ar, ^{37}Ar, and ^{36}Ar. **Moles $^{40}Ar_{rad}$:** absolute number of moles of ^{40}Ar measured in each diffusion step. $^{40}Ar_{rad}$ **(%):** percentage of radiogenic ^{40}Ar measured in each step. $^{39}Ar_{Ca}$ **(%)** and $^{36}Ar_{Ca}$ **(%):** percentages of ^{39}Ar and ^{36}Ar correcting those isotope abundances for calcium interference for each step. **K/Ca:** potassium/calcium ratio calculated for each step from the ^{39}Ar and ^{37}Ar isotope measurements. ^{39}Ar **(%):** percentage of ^{39}Ar released (100% total) during each step. **Age±s.d.:** age and standard deviation in age, in millions of years (Ma), calculated for each step]

NRF 7P (440 ft depth) [J=0.00036836]

plateau age = 0.884 ± 0.053 Ma; isochron age = 0.878 ± 0.242 Ma; inverse isochron age = 0.875 ± 0.211 Ma; total gas age = 0.724 ± 0.079 Ma

Temp (C°)	$^{40}Ar/^{39}Ar$	$^{37}Ar/^{39}Ar$	$^{36}Ar/^{39}Ar$	Moles$^{40}Ar_{rad}$	$^{40}Ar_{rad}$ (%)	$^{39}Ar_{Ca}$ (%)	$^{36}Ar_{Ca}$ (%)	K/Ca	^{39}Ar (%)	Age±s.d. (Ma)
450	48.00	3.740	0.16095	1.677^{-14}	1.4	0.24	0.48	0.131	48.4	0.500 ± 0.147
500	33.52	7.246	0.10990	1.985^{-14}	4.4	0.47	1.4	0.067	25.6	1.117 ± 0.105
550	22.63	13.692	0.07551	6.071^{-15}	5.0	0.88	3.7	0.035	10.1	0.864 ± 0.090
600	20.06	17.674	0.06824	3.391^{-15}	4.8	1.1	5.3	0.027	6.7	0.732 ± 0.108
650	30.09	17.631	0.10146	2.090^{-15}	3.9	1.1	3.6	0.027	3.4	0.892 ± 0.164
700	56.23	26.15	0.19286	8.484^{-16}	1.5	1.7	2.8	0.0184	2.0	0.624 ± 0.284
750	77.76	96.11	0.2818	3.859^{-16}	0.4	6.2	7.0	0.0048	2.1	0.264 ± 0.493
800	54.02	198.98	0.2348	$-1.445e^{-15}$	-6.1	12.9	17.4	0.0021	0.7	2.85 ± 1.06
900	43.63	150.79	0.17210	5.432^{-16}	4.4	9.7	18.0	0.0029	0.5	1.590 ± 1.091
1000	42.60	126.01	0.14899	1.659^{-15}	14.6	8.1	17.3	0.0036	0.5	5.07 ± 1.06

NRF 7P (486 ft depth) [J=0.00035049]

plateau age = 1.176 ± 0.027 Ma; isochron age = 1.178 ± 0.168 Ma; inverse isochron age = 1.181 ± 0.154 Ma; total gas age = 1.204 ± 0.031 Ma

Temp (C°)	$^{40}Ar/^{39}Ar$	$^{37}Ar/^{39}Ar$	$^{36}Ar/^{39}Ar$	Moles$^{40}Ar_{rad}$	$^{40}Ar_{rad}$ (%)	$^{39}Ar_{Ca}$ (%)	$^{36}Ar_{Ca}$ (%)	K/Ca	^{39}Ar (%)	Age±s.d. (Ma)
410	98.15	41.46	0.3833	-1.589^{-16}	11.9	2.9	3.0	0.0115	0.1	7.62 ± 7.63
450	52.74	13.935	0.16906	7.994^{-16}	7.5	0.98	2.3	0.035	1.0	2.51 ± 0.47
500	41.53	9.121	0.13580	4.002^{-15}	5.2	0.64	1.9	0.053	8.9	1.375 ± 0.108
525	37.27	8.054	0.12288	2.958^{-15}	4.4	0.57	1.8	0.060	8.8	1.032 ± 0.098
550	33.41	8.646	0.10948	4.899^{-15}	5.3	0.61	2.2	0.056	13.4	1.128 ± 0.087
575	29.24	9.816	0.09430	5.290^{-15}	7.5	0.69	2.9	0.050	11.7	1.391 ± 0.077
600	25.55	11.339	0.08210	4.568^{-15}	8.7	0.80	3.9	0.043	9.9	1.418 ± 0.075
625	21.97	13.821	0.07265	2.878^{-15}	7.5	0.98	5.3	0.035	8.4	1.049 ± 0.076
650	20.31	16.534	0.06769	2.367^{-15}	8.3	1.2	6.9	0.029	6.8	1.073 ± 0.085
700	20.79	16.767	0.06916	2.696^{-15}	8.4	1.2	6.8	0.029	7.4	1.120 ± 0.081
750	23.10	19.316	0.07787	2.286^{-15}	7.3	1.4	7.0	0.025	6.5	1.083 ± 0.092
800	27.98	18.409	0.09377	1.910^{-15}	6.4	1.3	5.5	0.026	5.1	1.150 ± 0.112
850	34.19	18.507	0.11588	1.279^{-15}	4.3	1.3	4.5	0.026	4.1	0.952 ± 0.137
900	68.27	17.976	0.2312	9.933^{-16}	2.1	1.3	2.2	0.027	3.3	0.932 ± 0.208
950	124.34	42.95	0.4217	1.967^{-15}	2.7	3.0	2.9	0.0111	2.8	2.15 ± 0.34
1000	161.07	119.85	0.5745	3.471^{-16}	0.8	8.5	5.9	0.0037	1.2	0.869 ± 0.585
1100	126.00	244.1	0.5015	-1.598^{-16}	-1.5	17.2	13.7	0.0017	0.3	1.472 ± 1.665

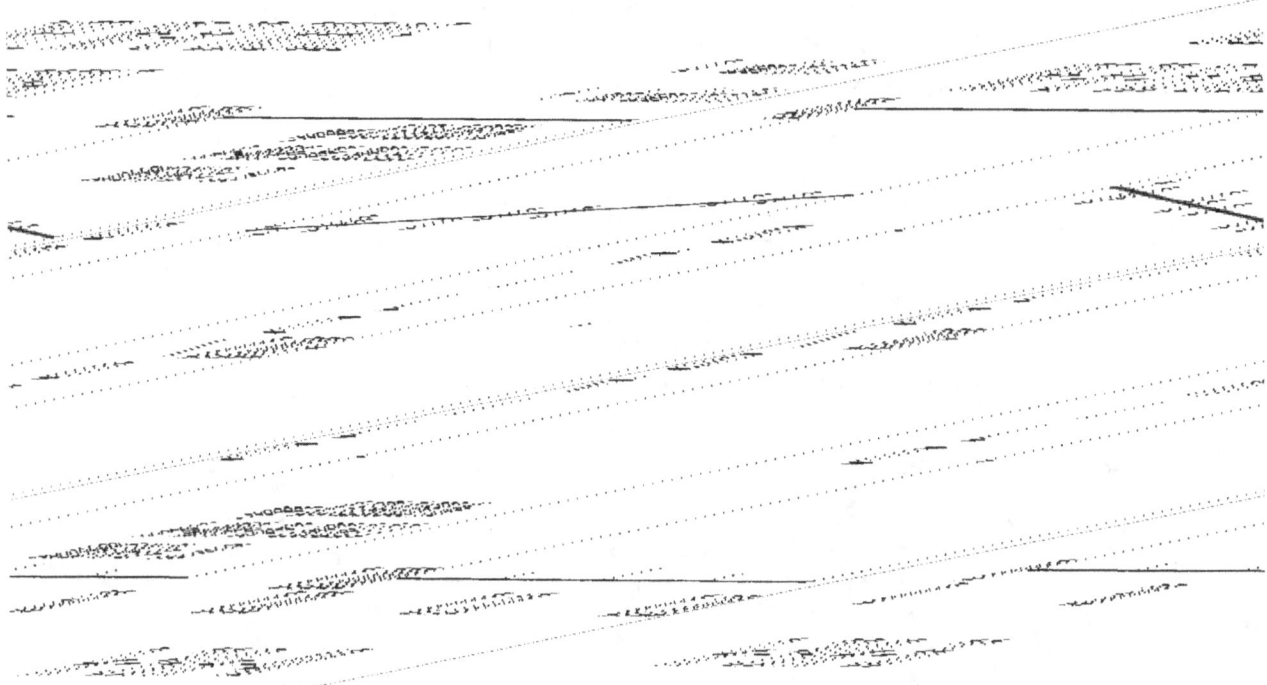

Figure B1-A.   Forward and inverse isochron diagrams for sample NRF-6P-155.

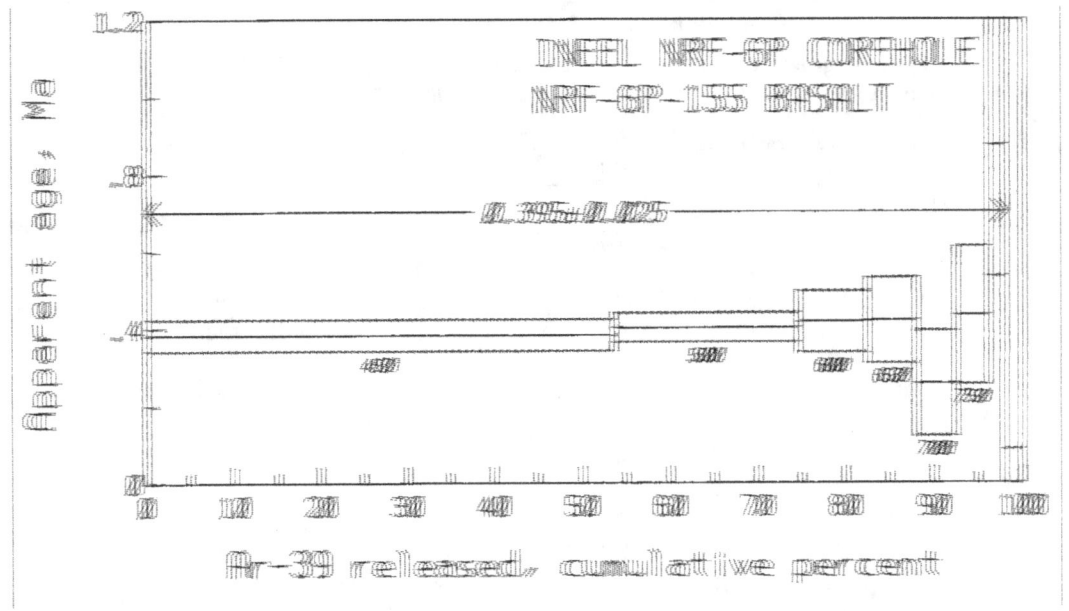

Figure B1-B.   Argon age plateau diagram for sample NRF-6P-155.

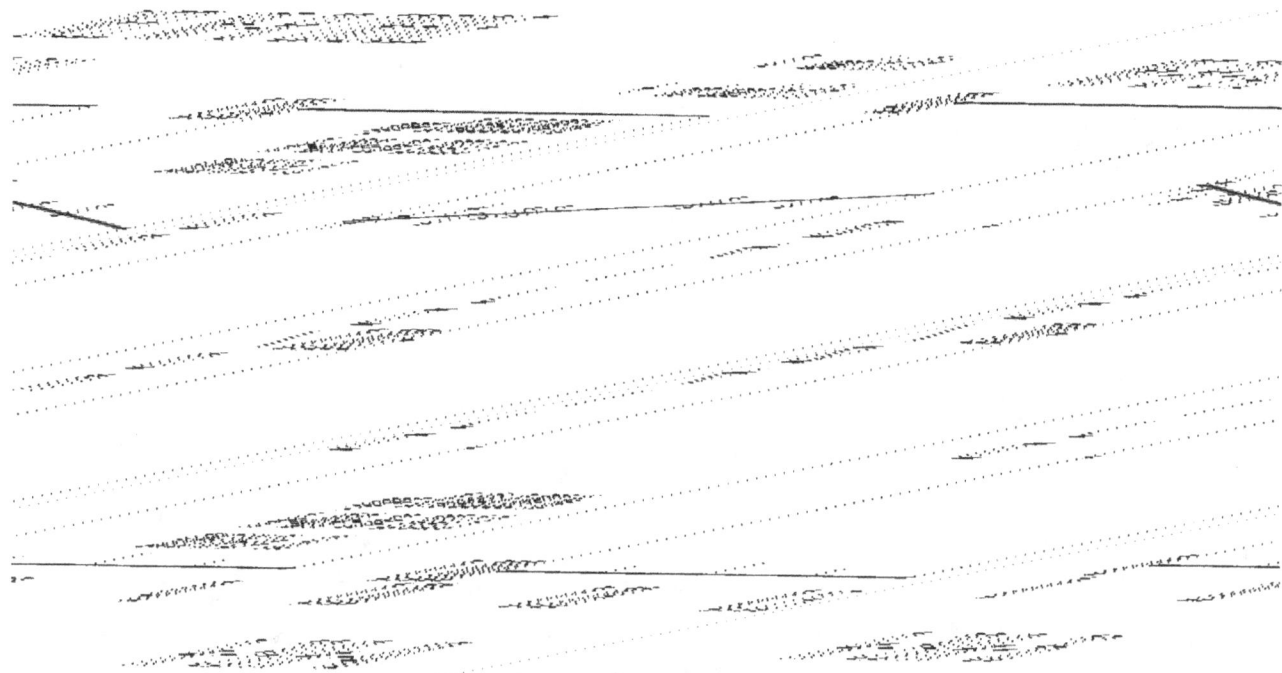

Figure B2-A. Forward and inverse isochron diagrams for sample NRF-6P-241.

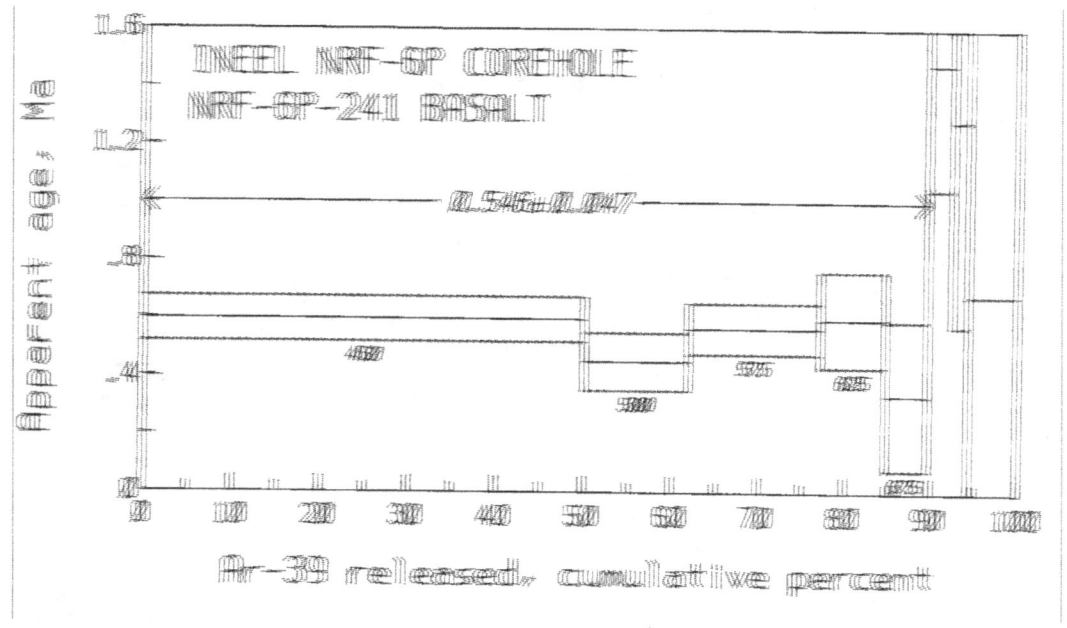

Figure B2-B. Argon age plateau diagram for sample NRF-6P-241.

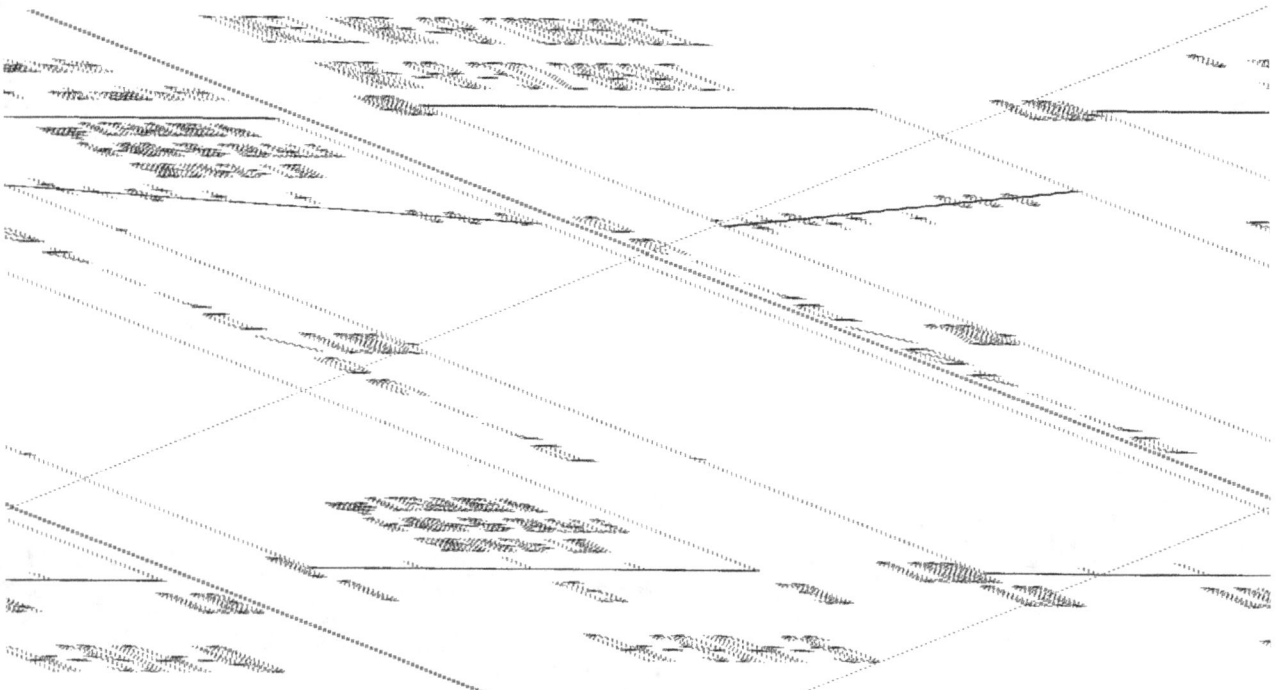

Figure B3-A.   Forward and inverse isochron diagrams for sample NRF-6P-372.

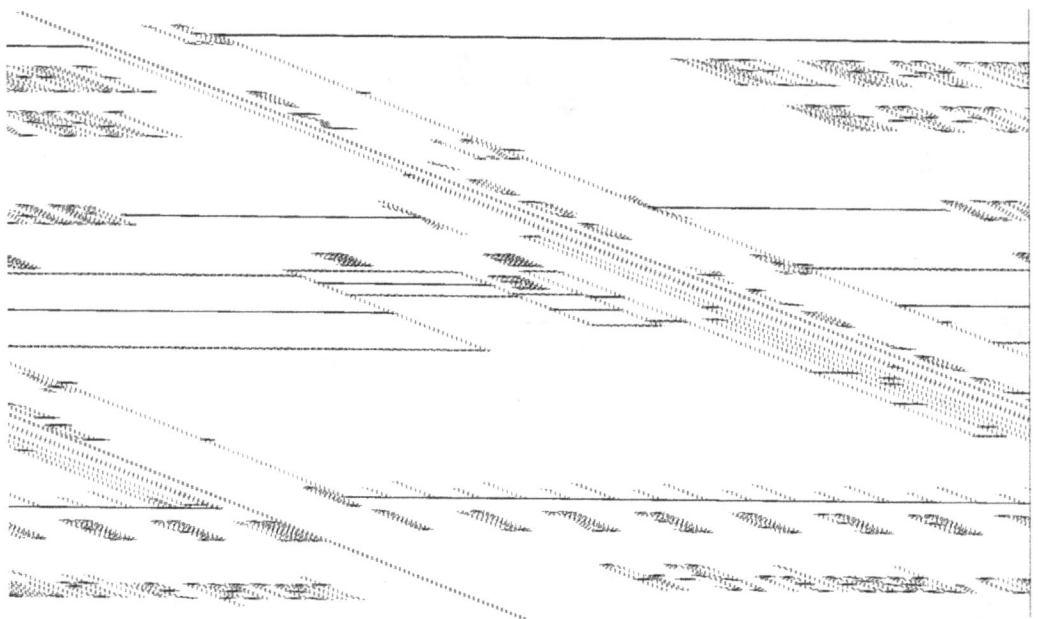

Figure B3-B.   Argon age plateau diagram for sample NRF-6P-372.

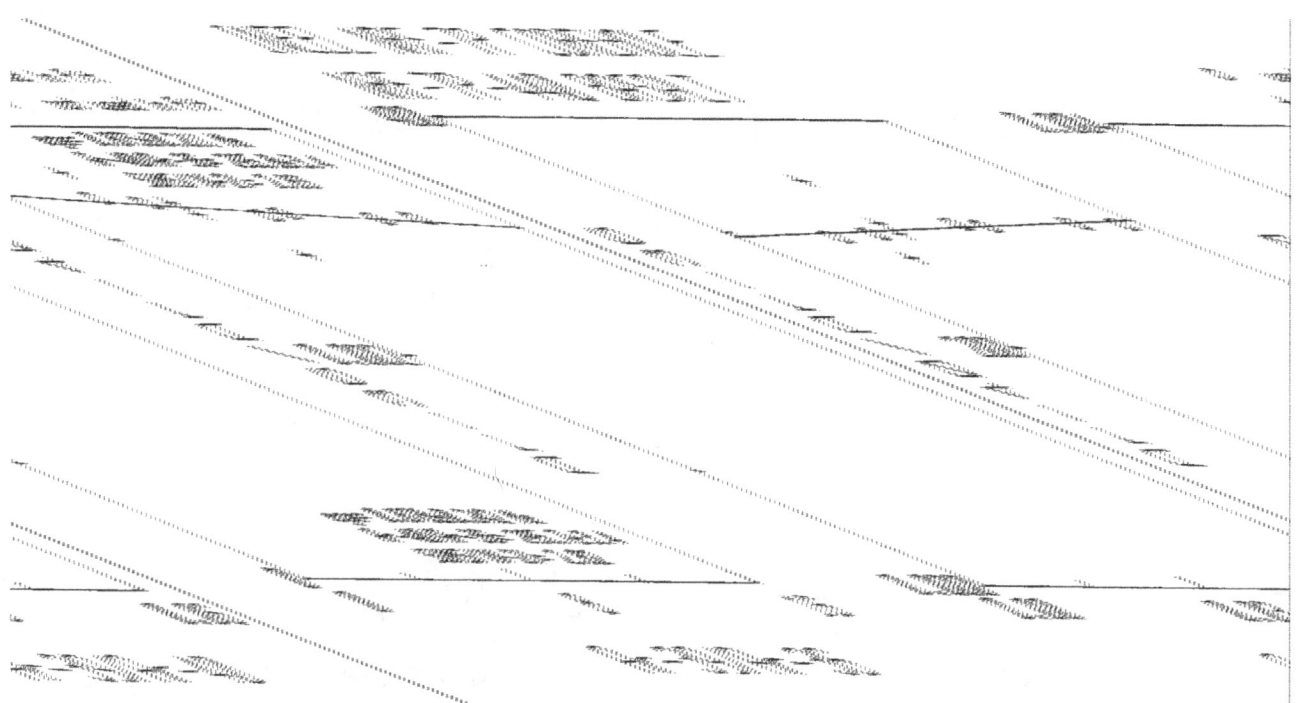

Figure B4-A. Forward and inverse isochron diagrams for sample NRF-7P-440.

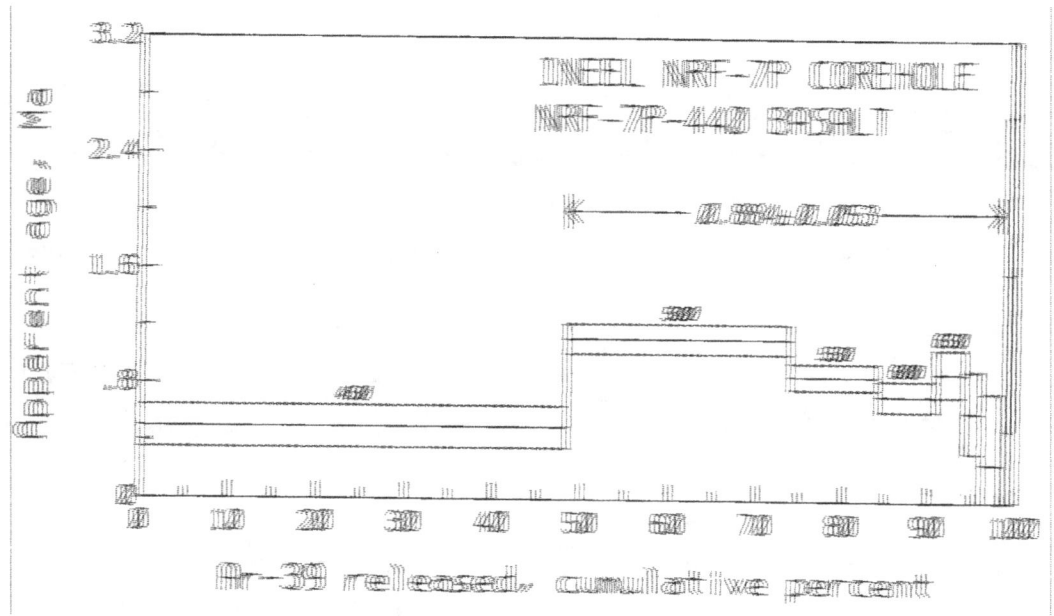

Figure B4-B. Argon age plateau diagram for sample NRF-7P-440.

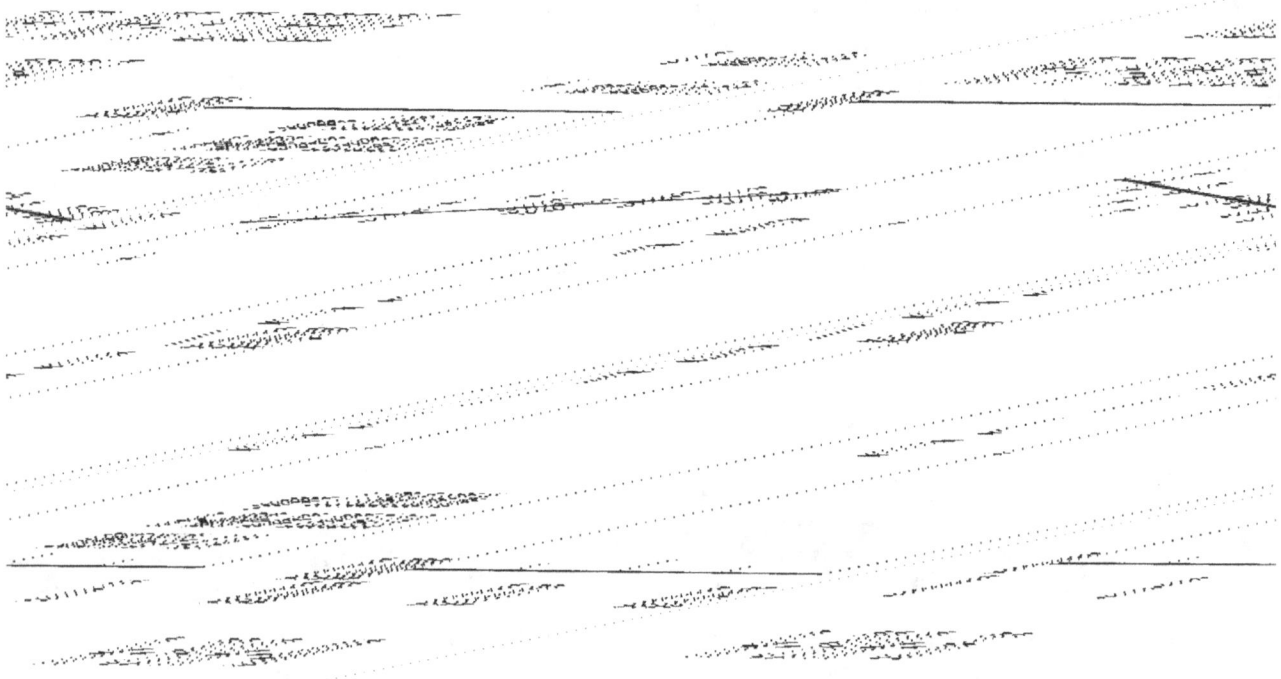

Figure B5-A.   Forward and inverse isochron diagrams for sample NRF-7P-486.

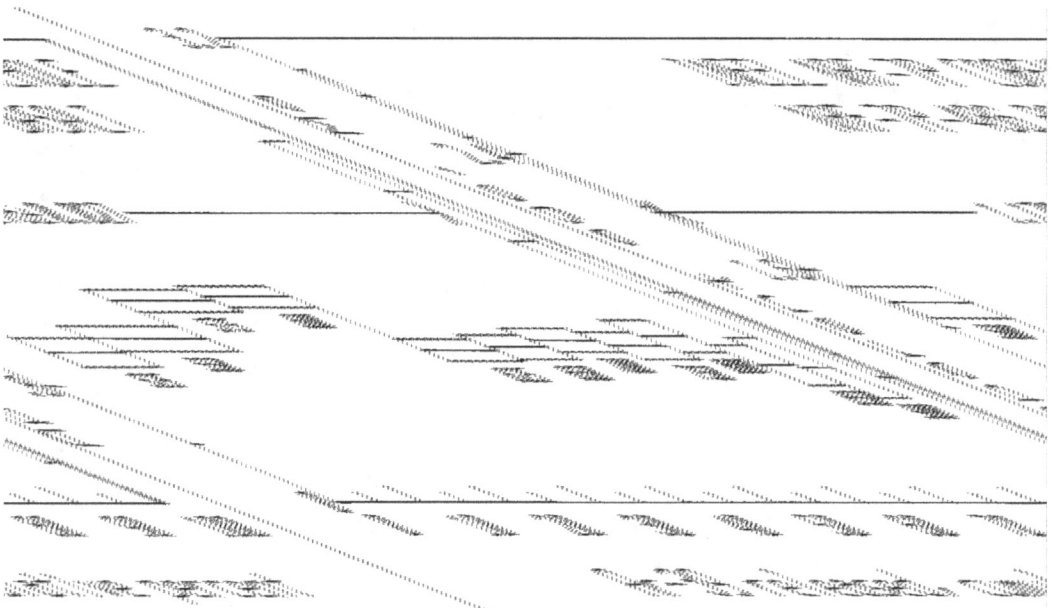

Figure B5-B.   Argon age plateau diagram for sample NRF-7P-486.

www.ingramcontent.com/pod-product-compliance
Lightning Source LLC
Chambersburg PA
CBHW081618170526
45166CB00009B/3023